T0140310

Fields Institute Monographs

VOLUME 38

The Fields Institute for Research in Mathematical Sciences

The Fields Institute is a centre for research in the mathematical sciences, located in Toronto, Canada. The Institutes mission is to advance global mathematical activity in the areas of research, education and innovation. The Fields Institute is supported by the Ontario Ministry of Training, Colleges and Universities, the Natural Sciences and Engineering Research Council of Canada, and seven Principal Sponsoring Universities in Ontario (Carleton, McMaster, Ottawa, Queen's, Toronto, Waterloo, Western and York), as well as by a growing list of Affiliate Universities in Canada, the U.S. and Europe, and several commercial and industrial partners.

More information about this series at http://www.springer.com/series/10502

Jérôme Losson • Michael C. Mackey
Richard Taylor • Marta Tyran-Kamińska

Density Evolution Under Delayed Dynamics

An Open Problem

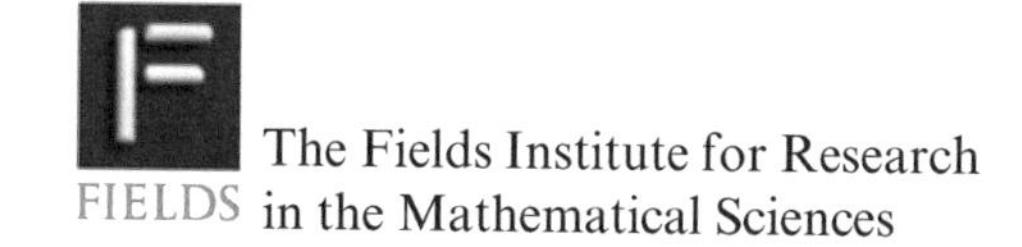

Jérôme Losson
BC Partners
London, UK

Michael C. Mackey
Department of Physiology
McGill University
Montreal, QC, Canada

Richard Taylor
Department of Mathematics and Statistics
Thompson Rivers University
Kamloops, BC, Canada

Marta Tyran-Kamińska
Institute of Mathematics
University of Silesia
Katowice, Poland

ISSN 1069-5273 ISSN 2194-3079 (electronic)
Fields Institute Monographs
ISBN 978-1-0716-1074-9 ISBN 978-1-0716-1072-5 (eBook)
https://doi.org/10.1007/978-1-0716-1072-5

This Springer imprint is published by the registered company Springer Science+Business Media, LLC, part of Springer Nature.
The registered company address is: 1 New York Plaza, New York, NY 10004, U.S.A.

Preface

This monograph has arisen out of a number of attempts spanning almost five decades to understand how one might examine the evolution of densities in systems whose dynamics are described by differential delay equations. Though we have no definitive solution to the problem, we offer this contribution in an attempt to define the problem as we see it and to sketch out several obvious attempts that have been suggested to solve the problem and which seem to have failed. We hope that by being available to the general mathematical community, they will inspire others to consider—and hopefully solve—the problem. Serious attempts have been made by all of the authors over the years and we have made reference to these where appropriate. Much of the material in this contribution has been previously published, but not all of it. A serious attempt has been made by Taylor (2004) and should be consulted for a variety of issues, some of which are repeated here. Along the same lines, many of the topics we raise here are also highlighted in Mitkowski and Mitkowski (2012), and in Mitkowski (2021).

The material is organized as follows:

Chapter 1 offers motivating examples to show why we want to study density evolution in systems with delayed dynamics.

In Chap. 2, we review briefly what is known about density evolution in systems with finite dimensional dynamics, starting with a description of the connection between dynamics and densities in Sect. 2.1. Section 2.3 reviews the situation for the commonly known situation in which the dynamics are described by ordinary differential equations. Section 2.4 briefly considers dynamics described by stochastic differential equations, while Sect. 2.5 does the same for finite-dimensional maps. Section 2.6 concludes with a description of the dynamic density evolution behaviors of ergodicity, mixing, exactness, and asymptotic periodicity.

Chapter 3 motivates the study of the dynamics of ensembles of differential delay equations through some simple numerical examples. Scction 3.1 relates the formal "density evolution" problem for differential delay equations to what is actually measured in an experimental setting. Section 3.2 gives numerical evidence for the existence of interesting ergodic properties of density evolution dynamics in the presence of delays. Chapter 4 considers the real mathematical problems involved ranging from the proper nature of the underlying space to the problem of defining a density and highlights all of the problems attendant in doing so. Chapter 5 outlines an approach that has been tried based on the Hopf functional. Section 5.1 introduces the notion of Hopf functionals, and Sect. 5.2 applies this to the specific case of delay differential equations.

Chapter 6 considers the problem reformulated as the method of steps. Finally, Chap. 7 considers the approximations to the delay problem, first examining a high-dimensional map approximation to the delay equation. Chapter 8 is devoted to developing approximate Liouville-like equations and an examination of invariant densities for differential delay equations. We conclude in Chap. 9.

We are indebted to numerous colleagues with whom we have discussed this problem over the years. We would like, in particular, to thank Andrzej Lasota (1932–2006), André Longtin, and Helmut Schwegler. The impetus for writing this is in large part due to a month-long workshop "Short Thematic Program on Delay Differential Equations" held at the Fields Institute (Toronto, May, 2015) and organized by Prof. Jianhong Wu (York University). Many colleagues there offered comments and suggestions, and for that we thank them, while other colleagues have generously shared their work with us ahead of publication.

This work was supported at various times by NATO, MITACS (Canada), the Humboldt Stiftung (Germany), the Natural Sciences and Engineering Research Council (NSERC, Canada), and the Polish NCN grant no. 2017/27/B/ST1/00100.

London, UK — Jérôme Losson

Montreal, QC, Canada — Michael C. Mackey

Kamloops, BC, Canada — Richard Taylor

Katowice, Poland — Marta Tyran-Kamińska

June 2020

Contents

Part I
Introduction and Background to Density Evolution Problems

Chapter 1 is an introductory section in which we set the stage for defining the problems of density evolution under the action of differential delay equations and motivating why we want to look at it.

Chapter 2 (page 9) talks briefly about density evolution in systems with finite-dimensional dynamics, starting with a description of the connection between dynamics and densities in Sect. 2.1. Section 2.3 reviews the situation for the commonly known situation in which the dynamics are described by ordinary differential equations. Section 2.4 briefly considers dynamics described by stochastic differential equations, while 2.5 does the same for finite-dimensional maps. This chapter concludes in Sect. 2.6 with a description of the dynamic density evolution behaviors of ergodicity, mixing, exactness, and asymptotic periodicity.

Chapter 1
Introduction and Motivation

In examining the dynamical behavior of a system there are fundamentally two options available to the experimentalist.

1. In the first option s/he will examine the dynamical trajectories of individuals, be they fundamental particles in a cloud chamber or cells in a petri dish or animals in an ecological experiment. In this case the experimentalist may be interested in replicating the experiment many times, and building up a statistical description of the observed behavior under the assumption (among others) that the trajectory behavior will be replicated between trials given the same initial conditions.
2. In the second option this approach will be forsaken for one in which the evolving statistics of large populations are examined. This is, of course, most familiar in statistical mechanics, but is also important in many other areas. The advantage of this approach is that if one can understand the dynamics of density evolution, then many interesting statistical quantities can be computed, and the results compared with experimental results.

Which approach is taken is sometimes a matter of choice, but often dictated by the nature of the individual units being studied as well as the types of experiments that are possible.

To illustrate these two points of view, consider the following, much-studied, example of a simple deterministic system whose evolution exhibits a species of random behavior. For a given real number x_0 (the "initial state" of the system) between 0 and 1, let x_1, x_2, x_3, etc. be defined by repeated application of the formula

$$x_{n+1} = 4x_n(1 - x_n), \quad n = 0, 1, 2, \ldots \tag{1.1}$$

One can view this formula as prescribing the evolution of the state of the system, x_n, at discrete times $n = 0, 1, 2, \ldots$. The evolution of this system is deterministic, in that once the initial state is specified equation (1.1) uniquely determines the

J. Losson et al., *Density Evolution Under Delayed Dynamics*, Fields Institute Monographs 38, https://doi.org/10.1007/978-1-0716-1072-5_1

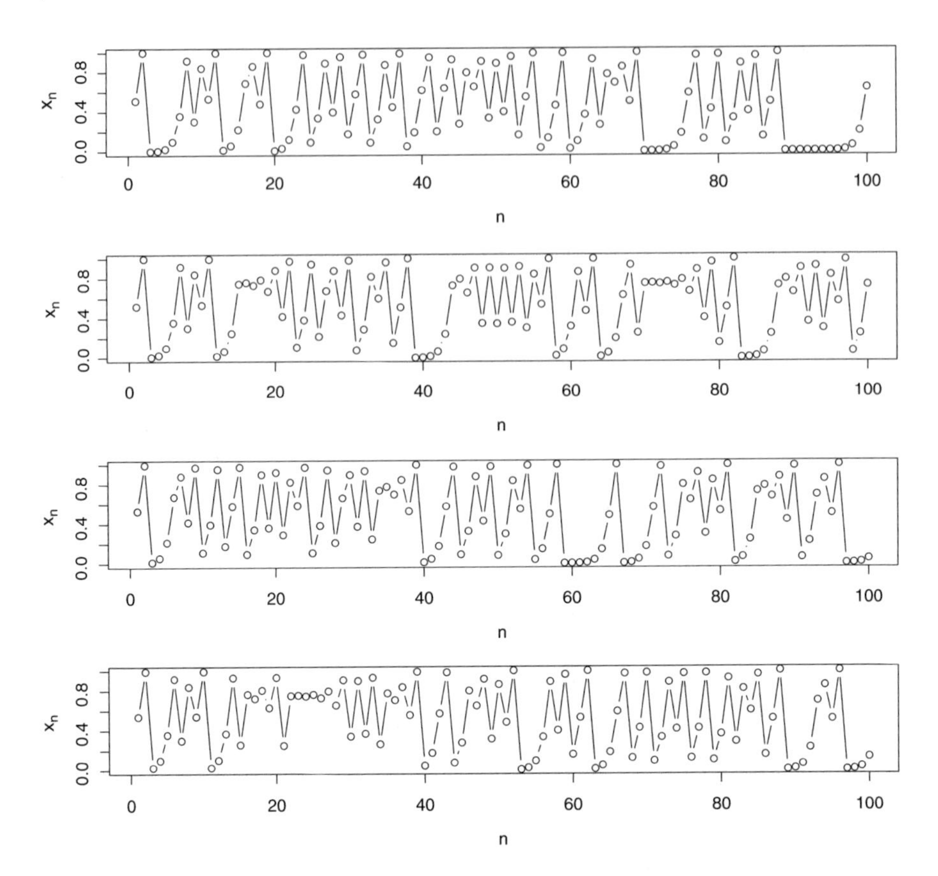

Fig. 1.1 Numerical trajectories for the map $x \mapsto 4x(1-x)$. The initial conditions differ only slightly for each trajectory

sequence of values $\{x_0, x_1, x_2, \ldots\}$ (i.e., the *trajectory* of the system) for all time. Thus, for example if $x_0 = 0.51$ we obtain

$$x_1 = .9996, x_2 \approx .0026, x_3 \approx .0064, x_4 \approx .025, x_5 \approx .099, \text{ etc.}$$

The qualitative behavior of this system is most easily appreciated graphically, as in Fig. 1.1 which plots x_n vs. n for typical trajectories obtained for different choices of x_0. Each of these trajectories is erratic, and random in the sense that no regularity is apparent. Furthermore, it can be seen by comparing the graphs in Fig. 1.1 that two nearly identical initial states yield radically different time evolutions. This phenomenon, termed "sensitive dependence on initial conditions" (Lorenz 1963), imposes strong limits on the predictability of this system over long periods of time: a small error in the determination of the initial condition rapidly becomes amplified to the extent that reliable prediction of the future states of the system eventually becomes impossible. Thus, despite being entirely deterministic, trajectories of this simple system have some hallmarks of essentially *random* phenomena: their behavior is irregular and unpredictable.

The mechanisms underlying the random character of this system are reasonably well understood (see, e.g., Collet and Eckmann 1980), the key notion being sensitivity to initial conditions and its consequences. However, within this framework it is difficult to approach questions of the type "what is the asymptotic behavior of a *typical* trajectory of this system?" Indeed, the very nature of sensitivity to initial conditions would seem to preclude any notion of "typical" behavior, since even very similar initial conditions eventually lead to their own very particular, uncorrelated evolutions.

However, different conclusions are reached if one takes a probabilistic point of view. Suppose that instead of being precisely determined, the initial state x_0 has associated with it some uncertainty. In particular, suppose we know the initial probability density, f, giving the probabilities of all possible values that x_0 can take. Then it makes sense to ask, "what will be the probability density of x_1, the new state after one iteration of the map (1.1)?" A precise answer to this question can be found using analytical methods described in Chap. 2. For an approximate answer, it suffices to simulate a large ensemble of different initial states x_0 distributed according to f, evolve these states forward under the map (1.1), and approximate the transformed density of the ensemble by constructing a histogram of the ensemble of values x_1. One can then proceed, in the same fashion, to determine the probability densities of subsequent states x_2, x_3, etc. Thus, even if the initial state x_0 is not known precisely, it is at least possible to give a probabilistic description of the system's evolution in terms of the evolution of a probability density.

The graphs in Fig. 1.2 show a particular choice for the probability density f of the initial state x_0, together with the subsequent densities of states x_1, x_2, x_3, x_4, obtained by numerical simulation of an ensemble of 10^6 initial values distributed according to f, iterated forward under the map (1.1). The striking feature of this figure is that the sequence of densities rapidly approaches an *equilibrium* or *invariant density* that does not change under further iteration. Moreover, the invariant density appears to be unique. This is supported by Fig. 1.3, which shows how a different choice of initial density evolves toward the same equilibrium density as before.

A different but related statistical approach to this system is to focus on the statistics of a single trajectory. For a given initial state x_0, by iterating $x_{n+1} = 4x_n(1 - x_n)$ we obtain an arbitrarily long sequence $\{x_n\}$ like the one illustrated in Fig. 1.1. A histogram of this sequence reveals the long-term frequency with which the trajectory visits various parts of the interval $[0, 1]$. Figure 1.4 shows such a histogram, for a trajectory of length 10^6. Remarkably, this histogram reproduces the invariant density shown in Figs. 1.2 and 1.3, which arises in a different context. Moreover, the same histogram is obtained for almost *any* choice of initial state.[1]

---

[1]There are exceptions, such as $x_0 = 0$, that yield trajectories with different (periodic) asymptotic behavior. These exceptions are very rare: in fact they constitute a set of Lebesgue measure 0.

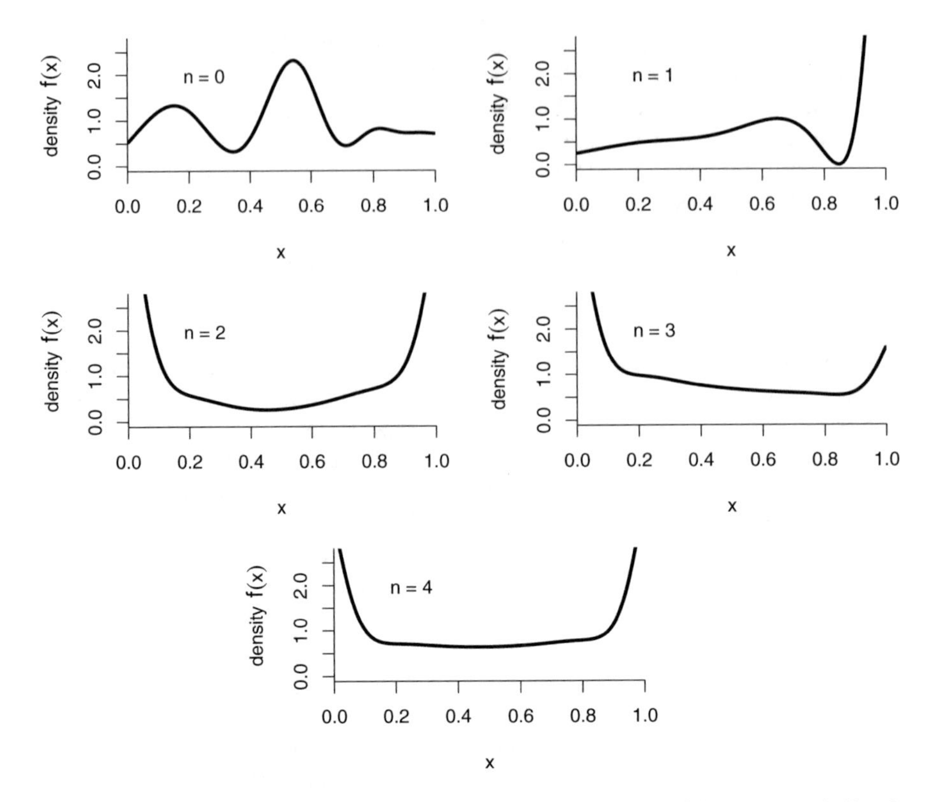

Fig. 1.2 Simulated evolution of an ensemble density f under iterations of the map $x \mapsto 4x(1-x)$

Thus, the invariant density describes the behavior of "typical" trajectories, i.e., those whose statistics are described by this particular probability density.

A probabilistic or ensemble treatment of dynamical systems provides a point of view complementary to one given in terms of the evolution of individual trajectories. The iterated map (1.1) is just one example of a system that behaves erratically on the level of individual trajectories, but has very regular asymptotic properties when considered at the level of probability densities. This observation appears to hold for many other systems. Moreover, it turns out that the converse holds as well: various regularity properties at the level of probability densities imply various degrees of disorder in the evolution of individual trajectories.

For a large class of systems in which the underlying dynamics are described by differential equations, or stochastic differential equations, or maps, there is a large *corpus* of methods that have been developed with which one can approach both of the types of data collection outlined above and the connection of that data to underlying dynamical systems theory.

However, many problems in the physical, and especially the biological, sciences involve the dynamic behavior of individual entities whose dynamics involve signif-

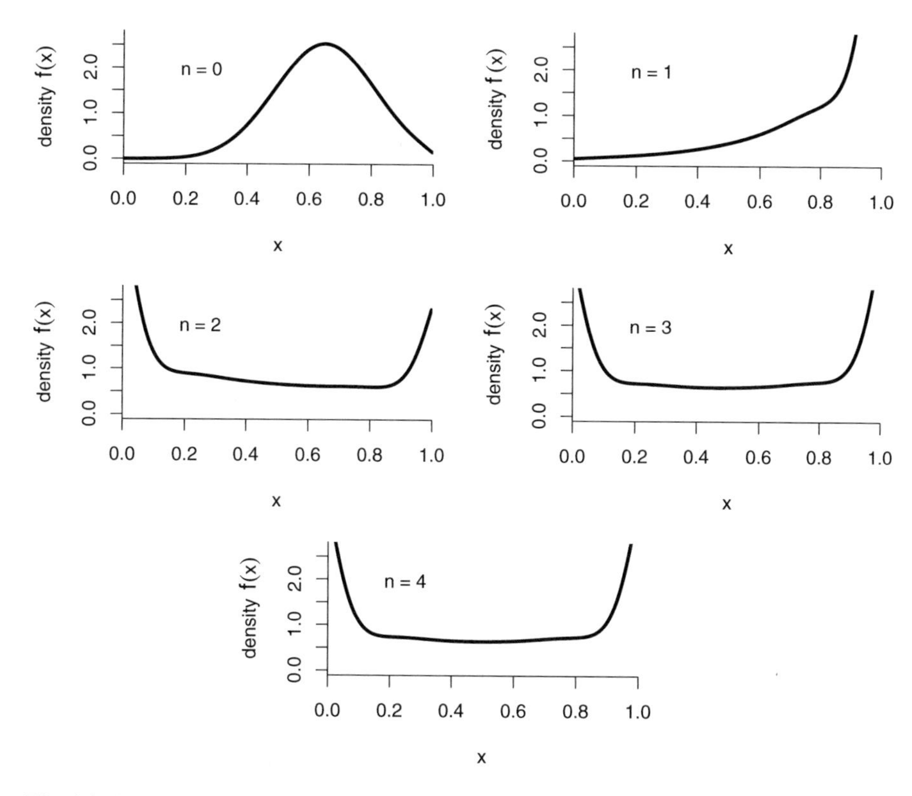

Fig. 1.3 As Fig. 1.2, with a different initial density

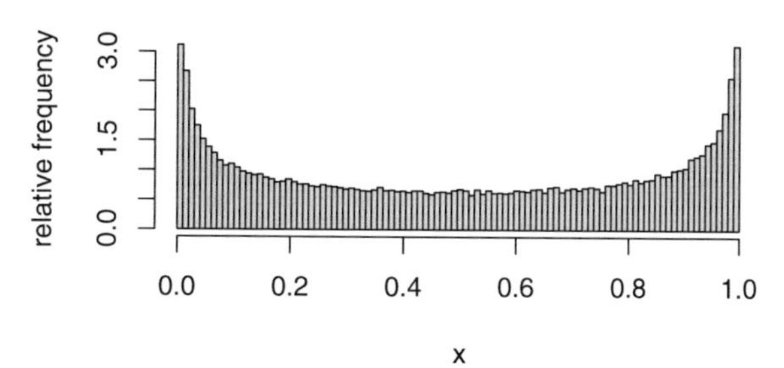

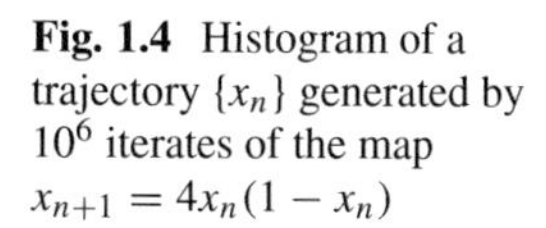

Fig. 1.4 Histogram of a trajectory $\{x_n\}$ generated by 10^6 iterates of the map $x_{n+1} = 4x_n(1 - x_n)$

icant delays. For problems like this, existing techniques to theoretically consider the evolution of densities are nonexistent. Repeated attempts to think of ways to formulate the evolution of densities in the presence of dynamics with delays have failed in even the most elementary respects (e.g., defining the fundamental mathematical aspects of the problem) and it is because of this failure that we present this Open Problem.

To be more concrete, if we have a variable x evolving under the action of some dynamics described by a differential delay equation

$$x'(t) = \epsilon^{-1}\mathcal{F}(x(t), x(t-\tau)), \qquad x(t) = \phi(t) \quad t \in [-\tau, 0], \tag{1.2}$$

then we would like to know how some initial density of the variable x will evolve in time, i.e., we would like to be able to write down an equation

$$\boxed{\text{UNKNOWN OPERATOR ACTING ON DENSITY} = 0.} \tag{1.3}$$

Unfortunately, we do not really know how to do this, and that is the whole point of this monograph. The reason that the problem is so difficult is embodied in Eq. (1.2) and the infinite-dimensional nature of the problem because of the necessity of specifying the *initial function* $\phi(t)$ for $t \in [-\tau, 0]$.

However, we do have some clues about what UNKNOWN OPERATOR should look like in various limiting cases. For example, if in Eq. (1.2), $\mathcal{F}(x(t), x(t-\tau)) = -x(t) + S(x(t-\tau))$ so (1.2) becomes

$$\epsilon x'(t) = -x(t) + S(x(t-\tau)), \qquad x(t) = \phi(t) \quad t \in [-\tau, 0], \tag{1.4}$$

then we expect that:

1. If $\tau \to 0$, then we should recover the normal Liouville operator (see Sect. 2.3 below) from UNKNOWN OPERATOR.
2. If we let $\epsilon \to 0$ and restrict t to $t \in \mathbb{N}$, then UNKNOWN OPERATOR should reduce to the Frobenius–Perron operator (see Sect. 2.5) for the map S.
3. If $\epsilon \to 0$, then from UNKNOWN OPERATOR we should recover the operator governing the evolution of densities *in a function space* under the action of the functional map.

$$x(t) = S(x(t-\tau)), \tag{1.5}$$

for $t \in \mathbb{R}^+$, though we do not know what that should be.

1.1 Summary

This chapter has introduced the concept of looking at dynamics via the evolution of densities as opposed to looking exclusively at the evolution of trajectories. We have highlighted, in a very qualitative and imprecise way, the unsolved problem that we are addressing through (1.3).

Chapter 2
Density Evolution in Systems with Finite-Dimensional Dynamics

For background material see Lasota and Mackey (1994).

2.1 Dynamics and Densities

In looking at ensemble behavior, the natural framework is to look at the evolution of a density as the description of the temporal behavior of the ensemble. Thus, we start by looking at the operators important for describing this density evolution.

We first start with a set X. Measure theorists often like to keep X pretty abstract, but for us X is going to be the *phase space* (more about this in the next section) on which all of our dynamics operates. Sometimes X will be a closed finite interval like $[0, 1]$, sometimes it may be $\mathbb{R}^+$, or even $\mathbb{R}^d$, and sometimes X is a function space. In any event, whatever X is we are going to assume that it does not have any pathological properties.

Let X be a space, $\mathcal{A}$ a σ-algebra, m a measure, and denote by $(X, \mathcal{A}, m)$ the corresponding σ-finite measure space. Let D be the subset of the space $L^1(X, \mathcal{A}, m)$ containing all densities, i.e. $D = \{f \in L^1 : f \geq 0, \|f\| = 1\}$. A linear map $P \colon L^1 \to L^1$ is a *Markov operator* if $P(D) \subset D$. If there is an $f_* \in D$ such that $Pf_* = f_*$ then f_* is called a *stationary density*.

In terms of dynamics, we consider $S_t \colon X \to X$, as time t changes. Time t may be either continuous ($t \in \mathbb{R}$) as, for example, it would be for a system whose dynamics were governed by a set of differential equations, or discrete (integer valued, $t \in \mathbb{Z}$) if the dynamics are determined by discrete time maps.

A *dynamical system* $\{S_t\}_{t\in\mathbb{R}}$ (or, alternately, $t \in \mathbb{Z}$ for discrete time systems) on a phase space X, is simply any group of transformations $S_t \colon X \to X$ having the properties $S_0(x) = x$ and $S_t(S_{t'}(x)) = S_{t+t'}(x)$ for $t, t' \in \mathbb{R}$ or $\mathbb{Z}$. Since, from the definition, for any $t \in \mathbb{R}$, we have $S_t(S_{-t}(x)) = x = S_{-t}(S_t(x))$, it is clear that

J. Losson et al., *Density Evolution Under Delayed Dynamics*, Fields Institute Monographs 38, https://doi.org/10.1007/978-1-0716-1072-5_2

dynamical systems are invertible, and solutions of systems of ordinary differential equations are examples of dynamical systems, called *flows*, as are invertible maps.

A *semi-dynamical system* or a *semi-flow* $\{S_t\}_{t\in\mathbb{R}^+}$ is any semigroup of transformations $S_t\colon X \to X$ such that $S_0(x) = x$ and $S_t(S_{t'}(x)) = S_{t+t'}(x)$ for $t, t' \in \mathbb{R}^+$ (or $\mathbb{N}$).

As one might guess, Markov operators are extremely valuable in terms of examining the evolution of densities under the action of a variety of dynamics, and in the next few sections we give concrete examples of these operators.

2.2 Frobenius–Perron Operator

Let $(X, \mathcal{A}, m)$ be a σ-finite measure space and $S_t\colon X \to X$ be a measurable[1] and nonsingular[2] transformation. Then the unique operator $P^t\colon L^1 \to L^1$ defined by

$$\int_A P^t f(x)\, m(dx) = \int_{S_t^{-1}(A)} f(x)\, m(dx) \tag{2.1}$$

is the *Frobenius–Perron operator* corresponding to S_t.

Thus, if f is a density, then Eq. (2.1) defining the Frobenius–Perron operator has an intuitive interpretation. Start with an initial density f and integrate this over a set B that will evolve into the set A under the action of the transformation S_t. However, the set B is $S_t^{-1}(A)$. This integrated quantity must be equal, since S_t is nonsingular, to the integral over the set A of the density obtained after one application of S_t to f. This final density is $P^t f$. Any density f_* for which $P^t f_* = f_*$ is called a *stationary density* of P^t.

The fact that the Frobenius–Perron operator is unique is a straightforward consequence of the Radon–Nikodym theorem. It is clear from the definition that the Frobenius–Perron operator is a Markov operator, and so P^t is a linear contracting operator. Also, if $f \geq 0$ then $P^t f \geq 0$ and $\|P^t f\| = \|f\|$. Finally it is easy to show that if $S_{nt} = S_t \circ \cdots \circ S_t$, and P^{nt} and P^t are, respectively, the Frobenius–Perron operator corresponding to S_{nt} and S_t, then $P^{nt} = P^t \circ \cdots \circ P^t = (P^t)^n$.

Sometimes the implicit definition Eq. (2.1) for the Frobenius–Perron operator allows one to obtain an explicit formula for P^t. For example, if $A = [a, x]$ and m is the Lebesgue measure then (2.1) becomes

$$\int_a^x P^t f(s)\, ds = \int_{S_t^{-1}([a,x])} f(s)\, ds$$

so

[1] $S_t^{-1}(A) \in \mathcal{A}$ for all $A \in \mathcal{A}$.

[2] $m(S_t^{-1}(A)) = 0$ for all $A \in \mathcal{A}$ such that $m(A) = 0$.

$$P^t f(x) = \frac{d}{dx} \int_{S_t^{-1}([a,x])} f(s)\, ds. \tag{2.2}$$

This process may be carried even further if the transformation is invertible so $S_t^{-1} = S_{-t}$ and S_{-t} has a continuous derivative with respect to x. Then, (2.2) becomes

$$P^t f(x) = f(S_{-t}(x)) \left| \frac{dS_{-t}(x)}{dx} \right| . \tag{2.3}$$

From this it is relatively straightforward to obtain a generalization of Eq. (2.3) valid for any invertible transformation S_t operating in $\mathbb{R}^d$. Namely

$$P^t f(x) = f(S_{-t}(x)) J^{-t}(x), \tag{2.4}$$

where $J^{-t}(x)$ is the absolute value of the determinant of the Jacobian matrix $DS_t^{-1}(x)$.

2.3 The Liouville Equation

Given a set of ordinary differential equations

$$\frac{dx_i}{dt} = \mathcal{F}_i(x), \qquad i = 1, \ldots, d \tag{2.5}$$

operating in a bounded region of $\mathbb{R}^d$, it is possible to derive an evolution equation for $P^t f_0(x)$ by using the invertibility of (2.5) in conjunction with (2.4). This gives the evolution equation for $f(x,t) \equiv P^t f_0(x)$:

$$\frac{\partial f}{\partial t} = -\sum_{i=1}^{d} \frac{\partial (f \mathcal{F}_i)}{\partial x_i},$$

which is known as the *generalized Liouville equation.*

2.4 The Fokker–Planck Equation

As an extension of the situation for ordinary differential equations, for stochastic differential equations of the form

$$dx = \mathcal{F}(x)dt + \sigma(x)dW(t),$$

where x is a d-dimensional vector and $W(t)$ is a standard Wiener process, the density $f(x,t) \equiv P^t f_0(x)$ satisfies the Fokker–Planck equation

$$\frac{\partial f}{\partial t} = -\sum_{i=1}^{d} \frac{\partial (f\mathcal{F}_i)}{\partial x_i} + \frac{1}{2}\sum_{i,j=1}^{d} \frac{\partial^2 (a_{ij} f)}{\partial x_i \partial x_j},$$

where $a_{ij}(x) = \sum_{k=1}^{d} \sigma_{ik}(x)\sigma_{jk}(x)$.

For references on the Fokker–Planck equation, there are literally scores to choose from. For people working in the area, most have their favorites and we find Gardiner (1983, 1991) and Risken (1984) to be especially useful.

2.5 Density Evolution in Maps

For the one-dimensional map

$$x_{t+1} = S(x_t)$$

the Frobenius–Perron operator is given by

$$P_S f(x) = \frac{d}{dx} \int_{S^{-1}([0,x])} f(u)\, du.$$

Example 2.1 For the tent map

$$S(x) = \begin{cases} 2x & \text{for } \ x \in \left[0, \frac{1}{2}\right) \\ 2(1-x) & \text{for } \ x \in \left[\frac{1}{2}, 1\right], \end{cases}$$

$S\colon [0,1] \to [0,1]$, and the corresponding Frobenius–Perron operator is given by

$$P_S f(x) = \frac{1}{2}\left[f\left(\frac{x}{2}\right) + f\left(1 - \frac{x}{2}\right)\right].$$

It is easily verified that the stationary density is $f_*(x) = 1_{[0,1]}(x)$ where $1_A(x)$ is the indicator function.

Further, for a one-dimensional map perturbed by a noise source ξ distributed with density g

$$x_{t+1} = S(x_t) + \sigma\xi_t,$$

the Markov operator governing the density evolution is given by

$$P_S f(x) = \int_X f(u) g(\sigma^{-1}(x - S(u)))\sigma^{-1} du.$$

2.6 The Dynamics of Density Evolution

As is the case when examining the temporal evolution of single trajectories emanating from a given initial condition in a dynamical system, there can be a variety of dynamical behaviors of densities when evolving from an initial density. The first ones that we consider are ergodicity, mixing, and asymptotic stability and all three of these can be characterized by the nature of the convergence of successive values of the densities.

The weakest type of convergence is contained in the property of ergodicity. Let $(X, \mathcal{A}, \mu)$ be a normalized measure space and $S\colon X \to X$ a non-singular transformation that preserves the measure μ[3] which has density f_*. S is *ergodic* if every invariant set $A \in \mathcal{A}$[4] is such that either $\mu(A) = 0$ or $\mu(X \setminus A) = 0$. Ergodicity is equivalent to the following.

Theorem 2.1 *S_t is ergodic with stationary density f_* operating in a phase space X if and only if for any integrable function g the time average of g along the trajectory of S_t is equal to the f_* weighted average of g over the entire phase space. That is,*

$$\lim_{t\to\infty} \frac{1}{t} \sum_{k=0}^{t-1} g(S_k) = \int_X f_*(x) g(x)\, m(dx) =: \langle f_*, g \rangle$$

in the discrete time case, or

$$\lim_{T\to\infty} \frac{1}{T} \int_0^T g(S_t(x))\, dt = \langle f_*, g \rangle$$

in the continuous time case.

Next in the hierarchy is the stronger property of mixing. Let $(X, \mathcal{A}, \mu)$ be a normalized measure space and $S\colon X \to X$ a transformation that preserves the measure μ which has density f_*. S is *mixing* if

$$\lim_{t\to\infty} \mu(A \cap S_{-t}(B)) = \mu(A)\mu(B) \quad \text{for all } A, B \in \mathcal{A}.$$

Mixing is equivalent to the following.

---

[3]The measure μ is invariant under A, i.e., $\mu(S^{-1}(A)) = \mu(A)$ for all $A \in \mathcal{A}$.

[4]$S^{-1}(A) = A$.

Theorem 2.2 *Let S_t be an ergodic transformation, with stationary density f_* of the associated Frobenius–Perron operator, operating in a phase space of finite f_* measure. Then S_t is mixing if and only if $\{P^t f\}$ is weakly convergent to f_* for all densities f, i.e.,*

$$\lim_{t\to\infty} \langle P^t f, g\rangle = \langle f_*, g\rangle$$

for every bounded measurable function g.

Finally, we have the strongest property of asymptotic stability. Let $(X, \mathcal{A}, \mu)$ be a normalized measure space, $S: X \to X$ a transformation that preserves the measure μ which has density f_*, and S such that $S(A) \in \mathcal{A}$ for each $A \in \mathcal{A}$. S is *asymptotically stable* if

$$\lim_{t\to\infty} \mu(S_t(A)) = 1 \quad \text{for all } A, B \in \mathcal{A}.$$

Asymptotic stability is equivalent to the following.

Theorem 2.3 *If S_t is an f_* measure preserving transformation and P^t is the associated Frobenius–Perron operator corresponding to S_t, then S_t is asymptotically stable if and only if*

$$\lim_{t\to\infty} \|P^t f - f_*\| = 0,$$

i.e., $\{P^t f\}$ is strongly convergent to f_*, for all initial densities f.

Remark 2.1 For an asymptotically stable transformation S with a stationary density f_*, we say that (S, f_*) is *exact*.

Remark 2.2 The three dynamic behaviors of densities we have examined are related in that asymptotic stability implies mixing which implies ergodicity. The converse is not true.

Remark 2.3 Ergodicity and mixing are properties that may be present in both dynamical and semi-dynamical systems. Asymptotic stability, however, is only possible in semi-dynamical systems.

A fourth type of known dynamic behavior that density evolution can display, that of asymptotic (or statistical) periodicity (Komorník and Lasota 1987; Komorník 1993), is as follows.

Theorem 2.4 *An ergodic transformation S is asymptotically periodic with period r if there exists a sequence of densities $g_1, \ldots, g_r$ and a sequence of bounded linear functionals $\lambda_1, \ldots, \lambda_r$ such that*

$$\lim_{t\to\infty} \|P^t(f - \sum_{j=1}^{r} \lambda_j(f) g_j)\| = 0. \tag{2.6}$$

The densities g_j have disjoint supports and $Pg_j = g_{\alpha(j)}$, where α is a permutation of $(1, \ldots, r)$. The invariant density is given by

$$g_* = \frac{1}{r}\sum_{j=1}^{r} g_j \tag{2.7}$$

and (S_r, g_j) is exact for every $j = 1, \ldots, r$.

Remark 2.4 Asymptotic periodicity is a density evolution property that may either be inherent in the dynamics (Provatas and Mackey 1991a; Losson and Mackey 1995) or induced by noise (Lasota and Mackey 1987; Provatas and Mackey 1991b) as in the next two examples.

Example 2.2 The generalized tent map on [0, 1] is defined by:

$$S(x) = \begin{cases} ax & \text{for} \quad x \in \left[0, \frac{1}{2}\right) \\ a(1-x) & \text{for} \quad x \in \left[\frac{1}{2}, 1\right]. \end{cases} \tag{2.8}$$

Ito et al. (1979a,b) have shown that the tent map Eq. (2.8) is ergodic, thus possessing a unique invariant density g_*. The form of g_* has been derived in the parameter window

$$a_{n+1} = 2^{1/2^{n+1}} < a \leq 2^{1/2^n} = a_n \qquad \text{for} \qquad n = 0, 1, 2, \cdots,$$

by Yoshida et al. (1983). Provatas and Mackey (1991a) have proved the asymptotic (statistical) periodicity of (2.8) with period $r = 2^n$, $n = 0, 1, \cdots$ for

$$2^{1/2^{n+1}} < a \leq 2^{1/2^n}.$$

Thus, for example, $\{P^t f\}$ has period 1 for $2^{1/2} < a \leq 2$, period 2 for $2^{1/4} < a \leq 2^{1/2}$, period 4 for $2^{1/8} < a \leq 2^{1/4}$, etc. Equation (2.8) is exact for $a = 2$.

The Frobenius–Perron operator corresponding to (2.8) is given by

$$Pf(x) = \frac{1}{a}\left[f\left(\frac{x}{a}\right) + f\left(1 - \frac{x}{a}\right)\right] 1_{[0, \frac{a}{2}]}(x).$$

Example 2.3 (Lasota and Mackey (1987) and Provatas and Mackey (1991b)) have studied the asymptotic periodicity induced by noise in a Keener map

$$S(x) = (ax + b) \pmod 1, \;\; 0 < a, b < 1$$

by studying the dynamics of

$$x_{n+1} = (ax_n + b + \xi_n) \pmod 1, \;\; 0 < a, b < 1, \tag{2.9}$$

when the noise source ξ is distributed with density g.

2.7 Summary

This chapter continues with the theme of Chap. 1 by introducing the Frobenius–Perron operator for the evolution of densities for ordinary differential equations, stochastic differential equations, maps, and stochastically perturbed maps. We then define various known types of dynamics of this density evolution (ergodic, mixing, asymptotically stable (exact), and asymptotically periodic).

Part II
Illustrating the Problem and Making It Precise for Differential Delay Equations

The following Chap. 3 (page 19) motivates the study of the dynamics of ensembles of differential delay equations through some simple numerical examples. Section 3.1 relates the formal "density evolution" problem for differential delay equations to what is actually measured in an experimental setting. Section 3.2 gives numerical evidence for the existence of interesting ergodic properties of density evolution dynamics in the presence of delays.

Chapter 4 (page 29) considers the real mathematical problems involved in understanding density evolution under delayed dynamics, ranging from the proper nature of the underlying space to the problem of defining a density, and highlights all of the problems attendant in doing so.

Chapter 3
Dynamics in Ensembles of Differential Delay Equations

3.1 What Do We Measure?

As pointed out in Chap. 1, there are fundamentally two types of data that are taken in experimental situations, and one is related to statistical properties of large ensembles of "units" that are typically assumed to have the same dynamics. If their dynamics are described by a differential delay equation of the form in Eq. (1.2), then we must consider what is likely to be measured. Figure 3.1 will aid in this.

In Fig. 3.1 we show a schematic depiction of what one would actually measure in an ensemble of units whose dynamic evolution is governed by a differential delay equation. We assume that there are N such units involved in our experiment, and that the experiment is started at time $t = 0$ with each of the N units having a history (= an initial function) on the interval $[-\tau, 0]$ preceding the start of the experiment. We let these N units evolve dynamically in time, and assume that we have a device able to record a histogram approximation to the density $f(x, t)$ of the distribution of the state variable x at time t.[1] Note that this measurement procedure is carried out at *successive* individual times and might be continuous.

Thus, what we *measure* is not unlike what we might measure in a system whose dynamics are evolving under the action of the system of ordinary differential equations (2.5). However, what we are able to *calculate* is far different.

Figure 3.2 illustrates the evolution of an ensemble of 100 solutions[2] of the so-called Mackey–Glass equation (Mackey and Glass 1977),

[1] It sometimes might be the case that we would not measure f, but rather might have estimates of various moments of f like $< x >$, $< x^2 >$, etc.

[2] Numerical solutions were computed here using the solver DDE23 (Shampine and Thompson 2001).

J. Losson et al., *Density Evolution Under Delayed Dynamics*, Fields Institute Monographs 38, https://doi.org/10.1007/978-1-0716-1072-5_3

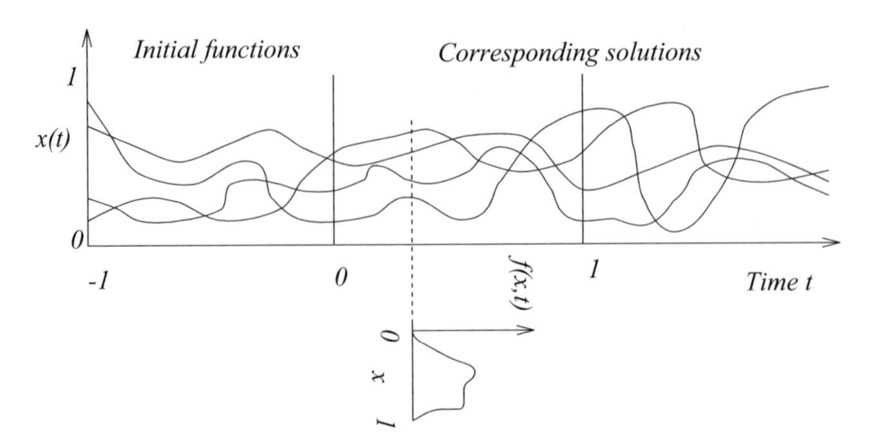

Fig. 3.1 A schematic illustration of the connection between the evolution of an ensemble of initial functions and what would be measured in a laboratory. In the case that the delay has been scaled to $\tau = 1$, an ensemble of N initial functions on $[-1, 0]$ is allowed to evolve forward in time under the action of the delayed dynamics. At time t we sample the distribution of the values of x across all N trajectories and form an approximation to a density $f(x, t)$. Taken from Losson and Mackey (1995) with permission

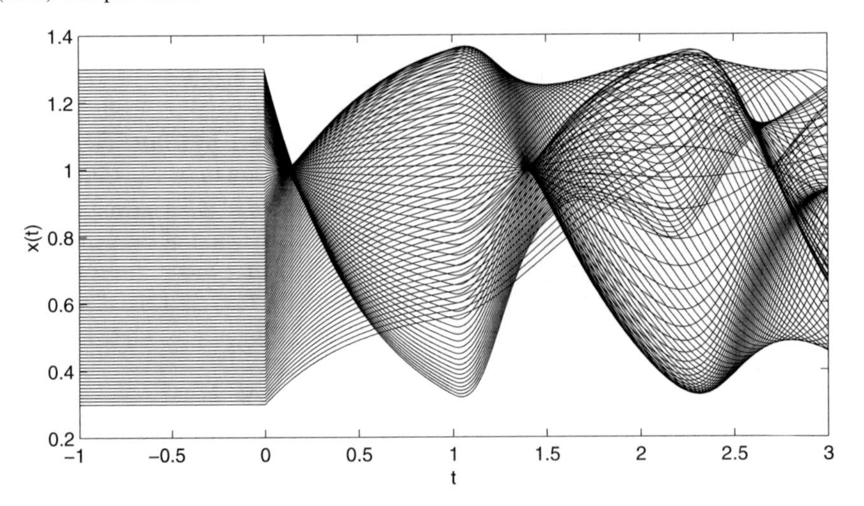

Fig. 3.2 An ensemble of 100 solutions of the Mackey–Glass equation (3.1), corresponding to an ensemble of constant initial functions with values uniformly distributed on the interval [0.3, 1.3]

$$x'(t) = -\alpha x(t) + \beta \frac{x(t-1)}{1 + x(t-1)^{10}},$$
$$\alpha = 2, \quad \beta = 4, \quad n = 10. \tag{3.1}$$

The parameter values chosen here correspond to the existence of a chaotic attractor. The solutions shown in Fig. 3.2 correspond to an ensemble of 100 constant initial

approaches a limiting density $f_*(x)$ as $t \to \infty$. That is, there appears to be an asymptotically stable invariant density for this system. The invariant density observed is in fact independent of the initial density.

Note that convergence to the invariant density is relatively slow, for example, compared to maps on the interval where statistical convergence occurs after only a few iterations of the Frobenius–Perron operator (cf. Figs. 1.2 and 1.3). The behavior seen here is not typical of dynamical systems considered elsewhere, and may have implications for the statistical mechanics of systems with delayed dynamics, for example the neural ensemble encoding mechanism proposed in Milton and Mackey (2000) where rapid statistical convergence plays an important role.

The brute force approach to densities has the tremendous advantage of being easy to implement—it requires only a method for numerically solving delay differential equations—and it is the obvious "quick and dirty" solution to the problem. However, it is a naïve approach, in that it provides no insight into the process by which $f(x,t)$ evolves. For example, the method offers only a heuristic explanation of the discontinuities and singularities that appear in Fig. 3.3. Moreover, constructing an accurate histogram can require millions of samples $\{x^{(i)}(t)\}$, hence millions of solutions of the differential delay equation must be computed. Especially when approximating the evolution of densities for large t, the amount of computation time necessary can render the method practically useless.

We now turn to two other examples investigated using a brute force method.

Losson and Mackey (1995) numerically studied the ensemble evolution of $N = 22{,}500$ differential delay equations of the form

$$\frac{dx}{dt} = -\alpha x + \begin{cases} a x_\tau & \text{if } x_\tau < 1/2 \\ a(1-x_\tau) & \text{if } x_\tau \geq 1/2 \end{cases} \qquad \frac{a}{\alpha} \in (1,2], \tag{3.2}$$

formed by considering the tent map (2.8) of Example 2.2 as the singular perturbation limit of the differential delay equation (1.4) with $\epsilon = 10$, and $a = 1.3$ in Eq. (2.8). Some of their results are shown in Fig. 3.6, clearly illustrating the existence of presumptive asymptotic periodicity in a continuous time setting that depends on the choice of the ensemble of initial functions.

Losson and Mackey (1995) have also numerically examined noise-induced asymptotic periodicity in a differential delay equation formed by considering the noisy Keener-map (2.9) as the singular perturbation limit of Eq. (1.4) where the noise source ξ is distributed with density g:

$$\frac{dx}{dt} = -\alpha x + [(a x_\tau + b + \xi) \mod 1] \quad 0 < a, b < 1. \tag{3.3}$$

The results of their simulations, shown in Fig. 3.7, give circumstantial evidence for the existence of noise induced asymptotic periodicity (Fig. 3.7c, d) as well as asymptotic stability (Fig. 3.7b).

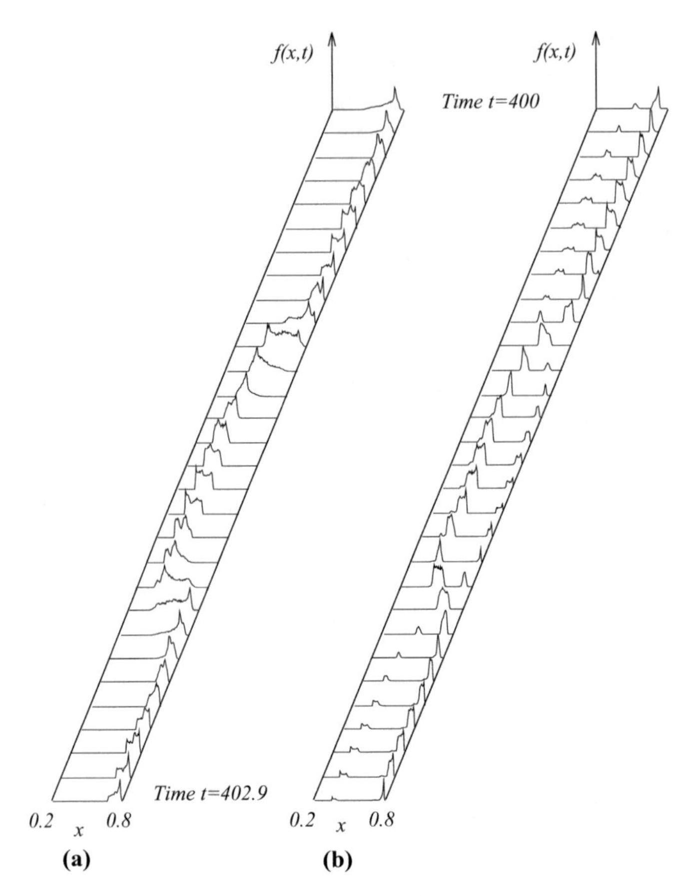

Fig. 3.6 Apparent asymptotic periodicity in the differential delay equation (3.2). The parameters are $a = 13$, $\alpha = 10$. Both (**a**) and (**b**) were produced with 22,500 initial functions. **(a)** Each of the initial functions was a random process supported uniformly on [0.65, 0.75]. **(b)** The initial functions were random processes supported either on [0.65, 0.75] (for 17,000 cases) or on [0.35, 0.45] (for the remaining 5500 initial functions). The cycling is not transient and is observed for all times. The dependence of the density cycle on the initial density reflects the dependence of the eventual form of the statistically periodic density sequence on the initial functions. See also Sect. 7.2. Taken from Losson and Mackey (1995) with permission

3.3 Summary

We have briefly illustrated, through more or less traditional numerical simulation, some of the apparently interesting types of density evolution that can occur in differential delay equations. Evidence is presented for apparent asymptotic stability (Fig. 3.5) as well as for inherent (Fig. 3.6) or noise-induced (Fig. 3.7) asymptotic periodicity. To do this we have used a "brute force" method of simulating a large ensemble of solutions for an ensemble of initial values chosen at random in

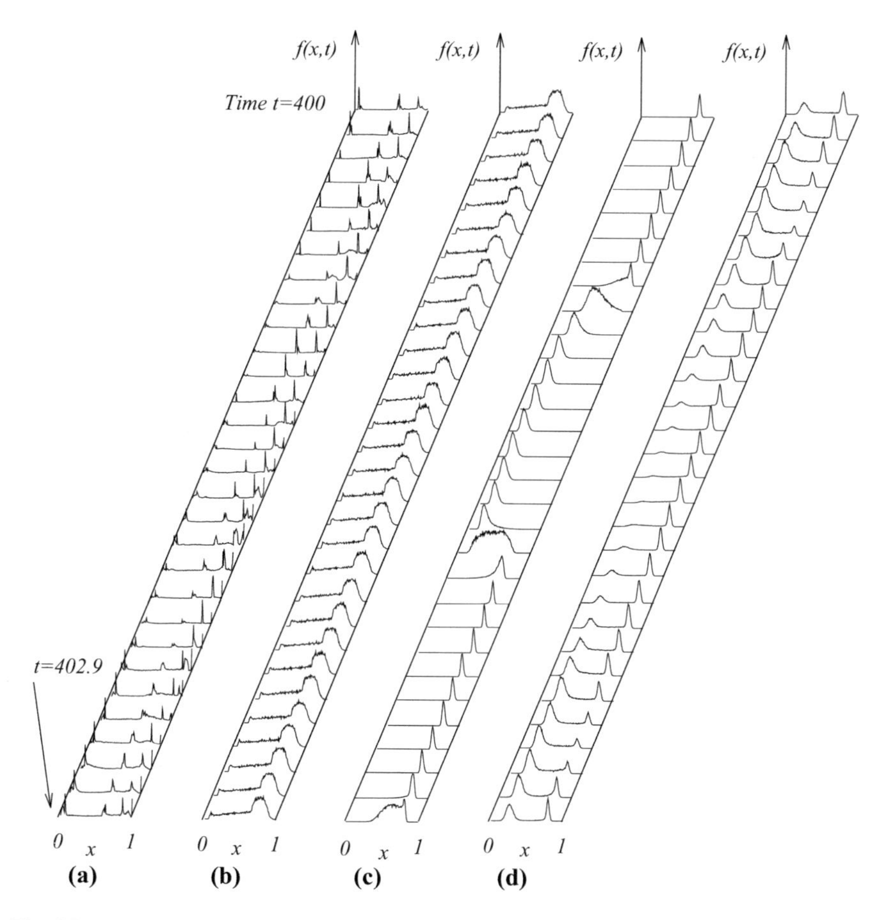

Fig. 3.7 Noise induced apparent asymptotic periodicity and stability in the stochastic differential delay equation (3.3). As in Fig. 3.6, each simulation was performed with 22,500 random initial functions. In all four panels, the parameters of the equation were $a = 0.5$, $b = 0.567$, $\alpha = 10$. For panels **(a)**–**(c)** the initial density was as in Fig. 3.6a. **(a)** No noise in the system: $f(x, t)$ is not a density, but a generalized function. **(b)** Noise supported uniformly on $[0, 0.1]$. The system is asymptotically stable, and $r = 1$ in (2.7). **(c)** Noise uniformly supported on $[0, 0.2]$, and $r = 2$ in (2.7). **(d)** Same noise as in **(c)**, with an initial density as in Fig. 3.6b. From Losson and Mackey (1995) with permission

accordance with the initial density. Because this method relies on adequate statistical sampling to obtain accurate results, it is computationally intensive to the point of being impractical for many applications. Taylor (2004, Chapter 4) has examined several other possible numerical schemes but none seem to offer a better way of investigating these dynamics.

Chapter 4
The Problem

4.1 Introduction

The central aim of this work is to try to understand how to apply probabilistic concepts, e.g., from ergodic theory, to the dynamics of delay differential equations. Before such a project can proceed, a number of foundational questions must be addressed. For instance,

- In what sense can a differential delay equation be interpreted as a dynamical system, i.e., with a corresponding evolution semigroup?
- What is the phase space for such a system?
- What semigroup of transformations governs the phase space dynamics of a differential delay equation?

The dynamical systems approach to delay equations is well established and provides standard answers to these questions. This theory is discussed below. As it happens the phase space for a delay differential equation is infinite-dimensional, which complicates matters considerably.

An ergodic approach to delay differential equations will require a theory of probability in infinite-dimensional spaces. The elements of this theory are discussed in Sects. 4.4 and 4.5. Naturally, there are technical and interpretational difficulties with doing probability in infinite dimensions. Indeed, the available mathematical machinery seems to be inadequate to deal with some of the problems that arise.

4.2 Delay Differential Equations

Delay differential equations, which are representative of the more general class of *functional* differential equations (Hale and Lunel 2013), take a great variety of forms. Delay equations having multiple time delays, time-dependent delays,

J. Losson et al., *Density Evolution Under Delayed Dynamics*, Fields Institute Monographs 38, https://doi.org/10.1007/978-1-0716-1072-5_4

and even continuous distributions of delays all arise in mathematical models of evolutionary systems (Driver 1977). To simplify matters we will restrict our attention to delay equations of the form

$$x'(t) = \mathcal{F}\big(x(t), x(t-\tau)\big), \tag{4.1}$$

where $x(t) \in \mathbb{R}^n$, $\mathcal{F}\colon \mathbb{R}^n \times \mathbb{R}^n \to \mathbb{R}^n$, and $\tau > 0$ is a single fixed time delay. Despite this restriction, the class of delay equations of the form (4.1) provides more than a sufficient arena for the considerations that follow.

4.2.1 Definition of a Solution

By a solution of the differential delay equation (4.1) we mean the following: if for some $t_0 \in \mathbb{R}$ and $\beta > t_0$, the function $x\colon [t_0 - \tau, \beta] \to \mathbb{R}^n$ satisfies (4.1) for $t \in [t_0, \beta]$, then we say x is a solution of (4.1) on $[t_0 - \tau, \beta]$. If $\phi\colon [t_0 - \tau, t_0] \to \mathbb{R}^n$ and x is a solution that coincides with ϕ on $[t_0 - \tau, t_0]$, we say x is a solution through (t_0, ϕ).[1]

Because Eq. (4.1) is autonomous (i.e., the right-hand side does not depend explicitly on t), it is invariant under time translation. That is, if $x(\cdot)$ is a solution, then, for any $T \in \mathbb{R}$, $x(\cdot + T)$ is also a solution. Consequently, the choice of initial time t_0 is arbitrary, and for the sake of convenience we can take $t_0 = 0$. Let $C = C([-\tau, 0], \mathbb{R}^n)$ be the space of continuous functions from $[-\tau, 0]$ into $\mathbb{R}^n$. Then if $\phi \in C$ and $x\colon [-\tau, \beta] \to \mathbb{R}^n$, we say x is a solution of (4.1) with *initial function* ϕ, or simply a solution through ϕ, if x is a solution through $(0, \phi)$.

Our consideration of delay equations necessarily imposes some constraints that are entirely natural. In order that a given differential delay equation describes an evolutionary process at all, we require the existence of solutions, at least for some subset of initial functions $\phi \in C$. Moreover, since ergodic theory is concerned largely with asymptotic properties, we require *global* existence, i.e., existence of solutions on the entire interval $[-\tau, \infty)$. To ensure that the process is deterministic we require that, for given $\phi \in C$, the solution through ϕ should be unique.

These constraints are in fact met under fairly mild restrictions on the right-hand side of (4.1). The following section presents the basic results of this theory that we will require. For further details of the existence and uniqueness theory for delay equations, see for example Driver (1977) and Hale and Lunel (2013).

[1] The difficulty that arises if ϕ does not satisfy (4.1) at t_0 is avoided if x' is interpreted as a right-hand derivative.

4.2.2 *Method of Steps*

Existence and uniqueness for a given differential delay equation can sometimes be shown indirectly, by representing the differential delay equation as a sequence of ordinary differential equations. This approach, known as the method of steps (Driver 1977), also furnishes a method of finding explicit solutions and will be investigated further in Chap. 6.

The differential delay equation problem

$$\begin{aligned} x'(t) &= \mathcal{F}\big(x(t), x(t-\tau)\big), \quad t \geq 0 \\ x_0 &= \phi, \end{aligned} \tag{4.2}$$

when restricted to the interval $[0, \tau]$, becomes the *ordinary* differential equation

$$x'(t) = \mathcal{F}\big(x(t), x_0(t-\tau)\big) \equiv \mathcal{G}_0\big(t, x(t)\big), \quad t \in [0, \tau],$$

since $x_0 = \phi$ is a known function. Under suitable hypotheses on $\mathcal{G}_0$, existence and uniqueness of a solution of this equation (hence a solution of (4.2)) on $[0, \tau]$ can be established. Denoting this solution by x_1 and restricting equation (4.2) to the interval $[\tau, 2\tau]$, we obtain the ordinary differential equation

$$x'(t) = \mathcal{F}\big(x(t), x_1(t-\tau)\big) \equiv \mathcal{G}_1\big(t, x(t)\big), \quad t \in [\tau, 2\tau],$$

for which we can again establish existence and uniqueness of a solution x_2.

Proceeding inductively, considering equation (4.2) as an ordinary differential equation on a sequence of intervals $[n\tau, (n+1)\tau]$, it is sometimes possible to show existence and uniqueness of a solution of the differential delay equation on $[-\tau, \infty)$. This approach is especially simple if $\mathcal{F}\big(x(t), x(t-\tau)\big) = \mathcal{F}\big(x(t-\tau)\big)$ is independent of $x(t)$, since existence and uniqueness of x_{n+1} then require only integrability of x_n, hence almost-everywhere continuity of ϕ is sufficient to guarantee existence and uniqueness of a solution on $[-\tau, \infty)$.

4.3 Delay Equation as a Dynamical System

To make sense of the differential delay equation (4.1) as prescribing the evolution of a deterministic system, we require that for any ϕ in C, a solution x through ϕ exists and is unique on $[-\tau, \infty)$. We will also require that $x(t)$ depend continuously on ϕ. Thus, from now on we will simply assume that sufficient conditions are satisfied to guarantee that these constraints are met, and the generic differential delay equation "initial data problem" we consider is the following:

$$x'(t) = \mathcal{F}\big(x(t), x(t-\tau)\big), \quad t \geq 0$$
$$x(t) = \phi(t), \quad t \in [-\tau, 0], \tag{4.3}$$

where $\phi \in C = C([-\tau, 0], \mathbb{R}^n)$.

Since Eq. (4.3) specifies the evolution of a variable $x(t) \in \mathbb{R}^n$, it might seem that such a differential delay equation could be regarded simply as a dynamical system on $\mathbb{R}^n$. However, $x(t)$ alone is inadequate as a "phase point," since the initial value $x(0)$ does not provide sufficient information to determine a solution. Indeed, in order that the right-hand side $\mathcal{F}\big(x(t), x(t-\tau)\big)$ is well defined for all $t \in [0, \tau]$, initial data consisting of values of $x(t)$ for $t \in [-\tau, 0]$ must be supplied, as in (4.3).

In general, to determine a unique solution of (4.3) for all $t \geq T$, it is necessary and sufficient to know the retarded values of $x(t)$ for all t in the "delay interval" $[T-\tau, T]$. Thus, Eq. (4.1) can only be considered as a dynamical system if the phase point at time t contains information about the solution $x(t)$ on the entire interval $[t-\tau, t]$. That this is in fact sufficient to define a dynamical system corresponding to the initial value problem (4.3) is shown in Hale and Lunel (2013).

Let C be the Banach space of bounded continuous functions from $[-\tau, 0]$ into $\mathbb{R}^n$, supplied with the sup norm. For each $t \geq 0$ define a transformation $S_t : C \to C$ by

$$(S_t\phi)(s) \equiv x_t(s) = x(t+s), \quad s \in [-\tau, 0], \tag{4.4}$$

where $x(t)$ is the solution of (4.3). Then we have (cf. Hale and Lunel (2013)):

Theorem 4.1 *The family of transformations S_t, $t \geq 0$, defined by Eq. (4.4), is a semi-dynamical system on C. That is,*

1. $S_0\phi = \phi \quad \forall \phi \in C$,
2. $(S_t \circ S_{t'})\phi = S_{t+t'}\phi \quad \forall \phi \in C, \ t_1, t_2 \geq 0$,
3. $(t, \phi) \mapsto S_t(\phi)$ *is continuous.*

In terms of this evolution semigroup, the initial data problem (4.3) can be written as an abstract initial value problem,

$$\begin{cases} x_t = S_t(x_0) \\ x_0 = \phi. \end{cases}$$

We call the function x_t the "phase point" at time t of the corresponding differential delay equation (4.3). The trajectory

$$\{x_t = S_t\phi : t \geq 0\}$$

is a continuous curve in the function space C. The relationship of the differential delay equation solution $x(t)$ to this trajectory is simple and is given by

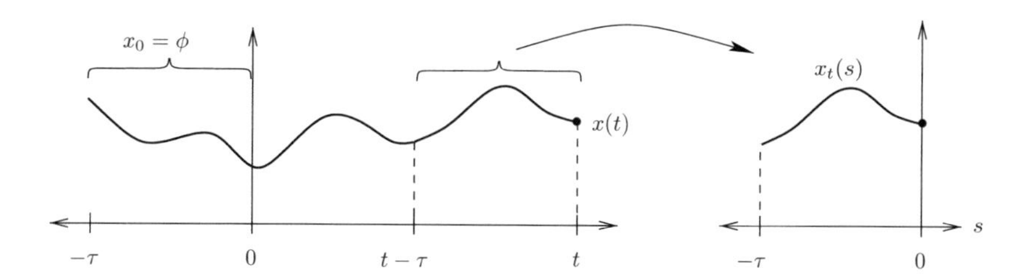

Fig. 4.1 Relationship between the solution $x(t)$ of the delay equation (4.3) and the phase point $x_t \in C$ of the corresponding dynamical system

$$x(t) = x_t(0).$$

That is, the solution $x(t)$ "reads off" the right endpoint of the phase point x_t. In other words, $x(t)$ can be interpreted as the projection of x_t under the map $\pi : C \to \mathbb{R}^n$ defined by $\pi(x_t) = x_t(0)$.

The action of S_t has a simple geometric interpretation. Since $(S_t\phi)(\cdot) = x(t + \cdot)$, S_t consists of a translation of the solution x followed by a restriction to the interval $[-\tau, 0]$. Figure 4.1 illustrates this action, together with the relationship of the state x_t to the differential delay equation solution $x(t)$.

The phase space of the dynamical system $\{S_t\}$ (and hence the phase space of the corresponding differential delay equation (4.3)), being the space of continuous functions on the interval $[-\tau, 0]$, is infinite-dimensional. The infinite-dimensionality of the phase space for delay equations complicates their analysis dramatically, and as we will see, it proves to be a serious barrier to developing a probabilistic treatment.

4.4 Frobenius–Perron Operator

Having determined how a delay differential equation defines a dynamical system, we are able to approach one of the fundamental problems posed in this monograph. That is, given a system whose evolution is determined by a differential delay equation (4.3) and whose initial phase point $\phi \in C$ is not known but is given instead by a probability distribution over all possible initial states, how does the probability distribution for the phase point evolve in time? Alternatively, we could consider the statistical formulation of the problem: given a large ensemble of independent systems, each governed by (4.3), and whose initial functions are distributed according to some density over C, how does this ensemble density evolve in time? It is of particular interest to characterize those probability distributions that are invariant under the action of the differential delay equation.

In a sense, the answer to this problem is simple and is provided by the Frobenius–Perron operator formalism, introduced in Chap. 2. Suppose the initial distribution of phase points is described by a probability measure μ on C. That is, the probability that the initial function ϕ is an element of a given set $A \subset C$ (correspondingly, the fraction of the ensemble whose initial functions are elements of A) is given by $\mu(A)$. Then, after evolution by time t, the new distribution is described by the measure ν given by

$$\nu = \mu \circ S_t^{-1}, \tag{4.5}$$

provided S_t is a measurable transformation on C (cf. Sect. 4.5.1). That is, after time t the probability that the phase point is an element of $A \subset C$ is $\nu(A) = \mu(S_t^{-1}(A))$. If the initial distribution of states u can be described by a density $f(u)$ with respect to some measure m, then after time t the density will have evolved to $P^t f$, where the Frobenius–Perron operator P^t corresponding to S_t is defined by

$$\int_A P^t f(u)\, m(du) = \int_{S_t^{-1}(A)} f(u)\, m(du) \quad \forall \text{ measurable } A \subset C. \tag{4.6}$$

Equations (4.5) and (4.6) might appear to answer the problem of the evolution of probability measures for delay differential equations. However, they amount only to a formal answer—essentially a symbolic restatement of the problem. In fact, everything that is specific to a given differential delay equation is contained in the symbol S_t^{-1}.

Although the differential delay equation can be expressed in terms of an evolution semigroup, there is no apparent way to invert the resulting transformation S_t. It is almost certain that such an inversion will be nontrivial, since solutions of delay equations frequently cannot be uniquely extended into the past (Driver 1977), so that S_t will not have a unique inverse. That is, S_t^{-1} may have numerous branches that need to be accounted for when evaluating $S_t^{-1}(A)$ in the Frobenius–Perron equation (4.6). This is a serious barrier to deriving a closed-form expression for the Frobenius–Perron operator P^t.

There are other subtle issues raised by Eqs. (4.5) and (4.6). The most apparent difficulty is that the integrals in (4.6) are over sets in a function space, and it is not apparent how such integrals can be carried out. More fundamentally, it is not clear what family of measures we are considering, and in particular what subsets $A \subset C$ are measurable (i.e., what is the relevant σ-algebra on C?). Also, in Eq. (4.6) what should be considered a natural choice for the measure m with respect to which probability densities are to be defined? For that matter, does it make sense to talk about probability densities over the function space C? These issues are explored in the following section.

4.5 Probability in Infinite-Dimensional Spaces

Any discussion of an ergodic theory of delay equations will require a theory of measure and integration on function spaces. In particular we need to discuss probability measures on the space C of continuous functions on the interval $[-\tau, 0]$, since this is a natural phase space for the differential delay equation (4.3). Colloquially speaking, we need to make precise the somewhat nonintuitive notion of selecting a *random function* from C.

Measure-theoretic probability provides a sufficiently abstract setting to accomplish this. We can represent a random variable $x \in X$ by its associated probability measure μ, with the interpretation that for a given subset $A \subset X$, $\mu(A)$ expresses the probability that $x \in A$. To ensure consistency with the axioms of probability, we cannot assign a probability to just any subset of X. Rather, μ must be defined on an appropriate σ-algebra—a collection of the so-called *measurable sets*. So choosing an appropriate σ-algebra on C is a necessary starting point.

4.5.1 Appropriate σ-Algebra

In real Euclidean spaces, the notion of measure derives from our physical intuition of length, area, volume, and their generalizations to higher dimensions. Thus, line segments in one dimension, and rectangles in two dimensions, are natural candidates for inclusion in the σ-algebras of choice for these spaces. The natural choice of σ-algebra would seem to be the smallest σ-algebra that contains all such sets—that is, the σ-algebra generated by these sets. This is the so-called *Borel σ-algebra*, which happens to coincide with the smallest σ-algebra that contains all open subsets.

A similar approach leads to a natural choice of σ-algebra for infinite-dimensional spaces such as C. That is, we take the Borel σ-algebra generated by the metric topology on C. With this choice, many important subsets of C become measurable, i.e., we can assign meaningful probabilities to them:

- any open set in C
- $\{u \in C : u(s) \in (a, b)\ \forall s \in [-\tau, 0]\}$; $a, b \in \mathbb{R}$ for $n = 1$
- any ϵ-ball $B_\epsilon(v) = \{u \in C : \|u - v\| < \epsilon\}$; $v \in C, \epsilon \in \mathbb{R}$.

Besides achieving the measurability of important sets for analysis, there is a more fundamental reason for choosing the Borel σ-algebra. Recall that studying the evolution of probability measures under a given transformation makes sense only if the transformation is *measurable*. Therefore, for our study of delay differential equations it is essential to choose a σ-algebra on which the semigroup S_t defined by Eq. (4.4) is measurable which is certainly true for the Borel σ-algebra.

However, it may be the case that the Borel σ-algebra on C is in fact not the most natural choice in the context of a probabilistic approach to delay differential equations. Certainly, as demonstrated in the following sections, measures on the Borel sets of infinite-dimensional spaces do not behave as we might like. However, in light of the preceding considerations, from now on we will consider only measures defined on the Borel sets of C.

4.5.2 Densities

Recall that if a measure μ is absolutely continuous with respect to a measure m, then it can be expressed as

$$\mu(A) = \int_A f\, dm, \tag{4.7}$$

where the integral is in the sense of Lebesgue, and $f \in L^1(X, m)$ is the density of μ with respect to m. Furthermore, any Lebesgue integrable function $f \in L^1(X, m)$ with

$$\int f\, dm = 1 \tag{4.8}$$

uniquely determines an absolutely continuous measure μ.

Since the relations (4.7) and (4.8) require only a σ-algebra and a measure m on X, they apply equally well in the more abstract setting of infinite-dimensional spaces such as C. That is, if C is equipped with a σ-algebra $\mathcal{A}$ and measure m on $\mathcal{A}$, then the function space $L^1(C) = L^1(C, \mathcal{A}, m)$ is unambiguously defined and any $f \in L^1(C)$ determines an absolutely continuous measure on C. However, in this context the intuitive appeal of densities is lacking: it is impossible to draw the graph of such a density. Even imagining a density on C seems beyond the power of one's imagination.

The analytical benefits of using densities also appear to be quite limited in infinite-dimensional spaces. The connection between measure theory and calculus in finite dimensions owes much to the theory of integration, notably the fundamental theorem of calculus and other theorems that facilitate calculations with integrals. There is no adequate theory of integration on function spaces that makes it possible to evaluate integrals like (4.7) on C (cf. comments in Losson and Mackey 1992). A notable exception to this is the Wiener measure, although this does not seem to be adequate for our purposes; see Sect. 4.5.4 (page 38).

Even assuming that a more powerful theory of integration may be available in the future, there remain some inherent difficulties with using densities to specify probability measures on infinite-dimensional spaces. Equation (4.6) for the evolution of a probability density f under the action of a semigroup S_t is valid only if S_t is nonsingular. That is, pre-images under S_t of m-measure-zero sets must have m-measure zero. It turns out to be difficult to guarantee this. In fact, on an infinite-dimensional space, every absolutely continuous measure fails to remain absolutely continuous under arbitrary translations (Yamasaki 1985). That is, for *any* measure m on C, there is some $v \in C$ for which the translation

$$T : u \mapsto u + v$$

is singular (in the measure-theoretic sense), and hence does not map densities to densities. If even translations do not lead to well-defined density evolution, there is little hope of studying delay equations with density functionals.

4.5.3 Lack of a "Natural" Measure on C

As if all of this did not complicate matters enough, if we are to work with densities on C there remains the problem of choosing a basic measure m with respect to which densities are to be defined (cf. Eq. (4.7)). This too turns out to be problematic.

In real Euclidean spaces we are accustomed to taking Lebesgue measure as the "natural" measure with respect to which densities are defined. That is, "a random number distributed uniformly on the interval [0, 1]" means "a random variable on [0, 1] distributed according to Lebesgue measure." Why is Lebesgue measure—of all possible measures—the gold standard for representing the concept of "uniformly distributed"?

The property of Lebesgue measure that selects it uniquely as the natural measure on Euclidean spaces is its translation invariance. Given a random variable x uniformly distributed on [0, 1], we expect that adding a constant a to x should result in a new random variable, $x + a$, that is uniformly distributed on $[a, a + 1]$, at least according to what seems to be the common intuitive notion of "uniformly distributed." More generally, a random variable uniformly distributed on any set in $\mathbb{R}^n$ should remain uniformly distributed if translated by a constant vector. Formally, the measure m on $\mathbb{R}^n$ that encapsulates uniform distribution should satisfy

$$m(A) = m(A + a), \quad \forall a \in \mathbb{R}^n, \ \forall \text{ measurable } A.$$

Another way to say this is that m is invariant under the translation group

$$T_a : x \mapsto x - a.$$

That is,

$$m = m \circ T_a^{-1} \quad \forall a \in \mathbb{R}^n. \tag{4.9}$$

Equation (4.9) uniquely defines the Borel measure m on the Borel σ-algebra on $\mathbb{R}^n$ (which agrees with Lebesgue measure on the Borel sets). This is a specific instance of Haar measure: every locally compact topological group (e.g., the translation group just considered on $\mathbb{R}^n$) has a unique group-invariant measure on the Borel σ-algebra, called the Haar measure, that is nonzero on any open set (Lang 1993, p. 313).

In light of these considerations, in choosing a natural measure on C it seems reasonable to seek a translation-invariant measure. After all, we would like that a uniformly distributed ensemble of functions in the unit ball in C should remain uniformly distributed under translation by any function in C. Unfortunately, the existence of a Haar measure on C is not guaranteed, since C is not locally compact.[2] In fact the situation is worse than that, as the following theorem demonstrates.

Theorem 4.2 (Hunt et al. (2002)) *Let X be an infinite-dimensional separable Banach space. If m is a nonzero translation-invariant measure on the Borel sets of X, then $m(A) = \infty$ for every open $A \subset X$.*

Since we expect any reasonable measure to be nonzero at least on some open sets, we can conclude that translation-invariance will not suffice to select a natural measure on C.

Aside from making the definition of densities on C ambiguous, the absence of a natural measure undermines one of the most important concepts in ergodic theory. A Sinai–Ruelle–Bowen (SRB; Young 2002) measure μ for a dynamical system S_t on X is one such that, for any $g \in L^1(X)$,

$$\lim_{T\to\infty} \frac{1}{T} \int_0^T g(S_t x)\, dt = \int g\, d\mu \tag{4.10}$$

for Lebesgue almost every x. Thus, the time average of g along almost every trajectory is equal to the spatial average of g weighted with respect to μ. Because $g(x)$ represents an arbitrary observable of the system, and μ encapsulates the asymptotic statistical behavior of $g(x(t))$ on almost every orbit of the system, it is widely accepted that an SRB measure is *the* relevant physical measure—the one that nature reveals to the experimentalist.

The notion of "almost every" in (4.10) is always unquestioningly taken to mean "Lebesgue almost every." As we have seen, for infinite-dimensional systems, and for delay equations in particular, we have no natural analog of "Lebesgue almost every," since there is no translation invariant measure to take the place of Lebesgue measure.

That this ambiguity emerges at all is somewhat amusing, since the notion of SRB measure was introduced on purely *physical* grounds. The very definition of SRB measure requires that we make precise the notion of "physically relevant"—but for delay differential equations this leads to considerations in the decidedly nonphysical setting of infinite-dimensional geometry, where it appears to be an inherently ambiguous term.

4.5.4 Wiener Measure

As noted, an adequate theory of integration on infinite-dimensional spaces is lacking. Such a theory is needed if we are to further develop the Frobenius–Perron

[2] A normed vector space is locally compact iff it is finite-dimensional (Lang 1993, p. 39).

operator formalism to characterize the evolution of densities for delay differential equations, which requires a theory of integration of functionals on the space C. This difficulty also arises in Chap. 5 (cf. Losson and Mackey 1992), in the context of a different approach to the evolution of densities for delay differential equations.

However, there is a notable exception worth mentioning. There is one probability measure (or family of measures) on a function space, called Wiener measure, for which there is a substantial theory of integration (Kac 1980). This measure plays an important role in quantum field theory (see, e.g., Ryder 1985) and is central to the theory of stochastic differential equations (Lasota and Mackey 1994, Ch. 11).

Let

$$C_0 = \{u \in C([0, 1], \mathbb{R}^n) : u(0) = 0\}.$$

A Brownian motion[3] is a stochastic process that generates a random path or "random function" $w \in C_0$ such that for a given $t \in [0, 1]$, $w(t)$ has Gaussian probability density (Losson and Mackey 1992)

$$f(x_1, \ldots, x_n) = \frac{1}{(\sqrt{2\pi t})^n} \exp\big[- (x_1^2 + \cdots + x_n^2)/(2t)\big].$$

Then, roughly speaking, Wiener measure μ_w assigns to a given subset $A \subset C_0$ a measure equal to the probability that a Brownian motion generates an element of A.

With Wiener measure it is possible to prove strong ergodic properties (e.g., exactness) for a certain class of partial differential equations (Brunovsky and Komornik 1984; Rudnicki 1985, 1987, 1988). The success of these investigations, together with the considerable machinery that has been developed around the Wiener measure, suggests that Wiener measure might be a good choice for the measure of integration in the study of other infinite-dimensional systems such as delay equations. However, in contrast with the quantum field equations and the partial differential equations mentioned above, the dynamical system $\{S_t\}$ corresponding to a delay equation does not leave the space C_0 invariant. That is, we cannot study S_t on C_0 alone. Thus, Wiener measure does not seem to be adequate for our purposes. Nevertheless, an approach based on Wiener measure might be still possible, and this suggests a potentially fruitful avenue for further investigation.

Before closing this section, at this juncture it is worth pointing out another technique that has been used in Rudnicki (2015) to examine the ergodic properties of infinite-dimensional dynamical systems. It is only applicable to linear semi-flows and involves the construction of an invariant mixing probability measure μ and then examining the covariance operator of this measure. If it can be shown that this operator is also the covariance operator of a Gaussian measure m, then m is an invariant mixing measure and the support of m is the closed linear subspace

[3]A Brownian motion is a continuous-time analog of a random walk starting at the origin. See, e.g., Lasota and Mackey (1994).

containing the support of the measure μ. This technique was used in Rudnicki (1993).

4.6 Conclusions

In this chapter we have examined a framework in which an ergodic treatment of delay differential equations might be developed. This provides a setting and terminology that will be needed for our discussion of possible avenues to study the ergodic properties of delay differential equations in subsequent chapters.

However, as far as the possibilities for the rigorous development of an ergodic theory of delay differential equations are concerned, this chapter also paints a rather pessimistic picture.

Equation (4.3) can be thought of as inducing a flow S_t on a phase space of continuous functions $C = C([-\tau, 0], \mathbb{R}^n)$, which is written as $x_t = S_t\phi$. By Theorem 4.1, $\{S_t : t \geq 0\} : C \to C$ is a strongly continuous semigroup. In one sense, it would seem that the evolution of a density under the action of this semigroup would be given by an extension of Eq. (2.1) given in (4.6). This writing of the evolution of the density f under the action of the semigroup of Frobenius–Perron operators $P^t : L^1(C) \to L^1(C)$ merely serves to highlight the major problems that we face.

The picture that emerges is a characterization of delay differential equations as infinite-dimensional dynamical systems on the phase space C of continuous functions on the interval $[-1, 0]$ (when the delay τ is normalized to 1). With this characterization, an ergodic theory of delay differential equations is possible in principle. In such a theory the mathematical objects of primary interest are probability measures on C. This entails having a theory of measure and probability on infinite-dimensional spaces. As we have seen, the foundations of this theory run aground on a number of technical and interpretational difficulties including the following:

- The phase space C of the dynamical system S_t is infinite-dimensional.
- Non-invertibility of the evolution semigroup $\{S_t\}$. How do we figure out what S_t^{-1} is?

  Typically, invertibility does not hold in differential delay equations, and the issue of non-invertibility in differential delay equations (also going under the name of backward continuation) is not as simple to deal with as in ordinary differential delay equations. Hale and Lunel (2013, Theorem 5.1, page 54) give very technical sufficient conditions for the existence of backward continuation which does not, in general, hold. For two very nice and understandable examples of the lack of invertibility in delay differential equations, see Driver (1977, Examples 3 and 5, Chapter V) and Sander et al. (2005).

- What family of measures m should we be considering in (4.6), and what subsets $A \subset C$ should we be thinking about? Also, the likely singularity of S_t with respect to most measures on C poses a problem.
- The integrals in (4.6) are over sets in a function space and we do not know how to carry out these integrations. Thus, there is a lack of an adequate theory of integration on infinite-dimensional spaces.
- Nonexistence of a natural (i.e., translation-invariant) measure on C.
- Ambiguity in the definition of SRB measure for infinite-dimensional systems.

Some of these difficulties (e.g., with integration in infinite dimensions) appear to require significant new mathematical tools. Others (e.g., with the choice of a natural measure on C and the definition of SRB measure) are simply ambiguities that arise when dynamical systems theory developed with only finite-dimensional systems in mind is carried over to an infinite-dimensional setting. There seems to be an absence of criteria by which these ambiguities could be resolved.

Clearly there are some significant obstacles in figuring out how to think about (4.6).

Remark It is extremely important to note the following.

In this chapter we have discussed the formidable and somewhat abstruse mathematical problems facing us in our quest for a theory of density evolution in systems with delayed dynamics, namely an infinite-dimensional phase space, the lack of a clearly useful measure, integration in infinite-dimensional spaces, and lack of a natural measure.

However, what is essential to remember is that when we collect experimental data from a system with delayed dynamics we are typically sampling a system as in Fig. 3.1 to produce a temporal sequence of density histograms like the single representation in that figure, or the sequence of density histograms as in Fig. 3.6 or in Fig. 3.7. The ultimate question is how to understand how to go from a "density in an infinite-dimensional space" to a sequence of apparently simpler evolving densities!

Part III
Possible Analytical Approaches

In this section we consider, in some detail, possible approaches that seem to be analytically sound.

Chapter 5 (page 45) outlines an approach that has been tried based on the Hopf functional. Section 5.1 introduces the notion of Hopf functionals, and the following Sect. 5.2 applies this to the specific case of delay differential equations.

Following that, in Chap. 6 (page 79) we outline a different approach that reformulates the problem using the method of steps.

Chapter 5
The Hopf Functional Approach

In this chapter we examine the possibility of extending techniques developed for the study of turbulent fluid flows to the statistical study of the dynamics of differential delay equations. Because the phase spaces of differential delay equations are infinite-dimensional, phase-space densities for these systems should be thought of as functionals.

The first part of the chapter is devoted to an exposition of Hopf functionals and some elementary considerations before turning to the second part of the chapter in which we derive a Hopf-like functional differential equation governing the evolution of these densities. The functional differential equation is reduced to an infinite chain of linear partial differential equations using perturbation theory. A necessary condition for a measure to be invariant under the action of a nonlinear differential delay equation is given. Finally, we show that the evolution equation for the density functional is the Fourier transform of an infinite-dimensional version of the Kramers–Moyal expansion.

5.1 Hopf Characteristic Functionals and Functional Differential Equations

This section briefly introduces a set of techniques to enable one to have a differential calculus for *functionals*, rather than just for the ordinary functions we usually consider, and shows how these may be used to study the behavior of infinite-dimensional systems like partial differential equations and differential delay equations.

J. Losson et al., *Density Evolution Under Delayed Dynamics*, Fields Institute Monographs 38, https://doi.org/10.1007/978-1-0716-1072-5_5

5.1.1 Functionals and Functional Derivatives

A *functional* may be considered a function of infinitely many variables. Thus, one could consider a function $h\colon \Delta \to R$ and then a *space of functions* $C(\Delta)$, so a functional Ψ is $\Psi\colon C(\Delta) \to R$. The functional $\Psi(h)$ should be thought of as a natural extension and generalization of a function $F(x)$ of many variables.

In the normal calculus we are accustomed to the definition of the partial derivative of a function of several variables, e.g., with $F\colon R^d \to R$ with $x = (x_1, \cdots, x_i, \cdots, x_d)$ and $\bar{x} = (\bar{x}_1, \cdots, \bar{x}_i, \cdots, \bar{x}_d)$ we have

$$\frac{\partial F(x)}{\partial x_i} = \lim_{\bar{x}_i \to x} \frac{F(\bar{x}) - F(x)}{\bar{x}_i - x_i}.$$

We want to develop a natural extension of this concept to functionals $\Psi(h)$. As in the more usual case we expect that this extension will:

1. Depend on Ψ; and
2. Depend on the choice of the "point" h.

How are we to proceed? Consider two functions h and a second one $\bar{h}$ that is just a bit different from h in some neighborhood of x_0. Then the "increase" of the functional $\Psi(h)$ in going from h to $\bar{h}$ is just

$$\Psi(\bar{h}) - \Psi(h).$$

Further, the area between h and $\bar{h}$, if $\bar{h} \geq h$, is just

$$\int_\Delta (\bar{h} - h)\, dx.$$

Thus, we take the expression

$$\lim_{\bar{h} \to h,\, \text{supp}(\bar{h}-h) \to x_0} \frac{\Psi(\bar{h}) - \Psi(h)}{\int_\Delta (\bar{h} - h)\, dx}$$

to be the definition of the *functional* or *Volterra derivative* of $\Psi(h)$ and we denote it by

$$\frac{\delta \Psi(h)}{\delta h(x_0)}.$$

Hence

$$\frac{\delta \Psi(h)}{\delta h(x_0)} = \lim_{\bar{h} \to h\ \text{supp}(\bar{h}-h) \to x_0} \frac{\Psi(\bar{h}) - \Psi(h)}{\int_\Delta (\bar{h} - h)\, dx}.$$

Instead of the above we might simply write the increase of Ψ between h and $\bar{h}$ as

$$\Psi(\bar{h}) - \Psi(h) = c \int_{\Delta} (\bar{h} - h)\, dx + O\left(\int_{\Delta} |\bar{h} - h|\, dx\right),$$

where

$$\lim_{\bar{h}\to h,\ \mathrm{supp}(\bar{h}-h)\to x_0} O\left(\int_{\Delta} |\bar{h} - h|\, dx\right) = 0,$$

and the convergence to zero is faster than that of

$$\int_{\Delta} (\bar{h} - h)\, dx.$$

In this case, the constant c is just taken to be the functional derivative, i.e.

$$c = \frac{\delta \Psi(h)}{\delta h(x_0)}.$$

In the ordinary calculus the quantity

$$\frac{\partial F(x)}{\partial x_i},$$

for a fixed i, may be considered as a new function and

$$\frac{\partial^2 F(x)}{\partial x_j \partial x_i}$$

calculated. Analogously, we can fix x_0 (analogous to fixing i) in the expression

$$\frac{\delta \Psi(h)}{\delta h(x_0)},$$

and then calculate

$$\frac{\delta^2 \Psi(h)}{\delta h(x_i) \delta h(x_0)}.$$

Example 5.1 (Functional Derivative of a Linear Functional) Having defined a functional derivative, let us try to calculate one. Specifically, let us try to calculate the functional derivative of the *linear functional*

$$\Psi(h) = \int_{\Delta} h(x)\psi(x)\, dx, \tag{5.1}$$

where $\psi(x)$ is taken to be continuous.

Now the increase of Ψ can be simply written as

$$\begin{aligned}\Psi(\bar{h}) - \Psi(h) &= \int_\Delta \psi(x)[\bar{h}(x) - h(x)]\,dx \\ &= \int_\Delta \psi(x_0)[\bar{h}(x) - h(x)]\,dx + \int_\Delta [\psi(x) - \psi(x_0)][\bar{h}(x) - h(x)]\,dx \\ &= \psi(x_0)\int_\Delta [\bar{h}(x) - h(x)]\,dx + \int_\Delta \epsilon|x - x_0|[\bar{h}(x) - h(x)]\,dx \\ &= \psi(x_0)\int_\Delta [\bar{h}(x) - h(x)]\,dx + O\left(\int_\Delta |\bar{h}(x) - h(x)|\,dx\right).\end{aligned}$$

However, this is just precisely the form we had in our alternative formulation of the functional derivative, and we thus conclude that for Eq. (5.1) the functional derivative is

$$\frac{\delta\Psi(h)}{\delta h(x_0)} = \psi(x_0), \tag{5.2}$$

i.e., it is just the kernel of (5.1) evaluated at x_0 and is clearly independent of h.

Remark 5.1 It is interesting to compare this result with the situation in the ordinary calculus where we differentiate a linear function

$$F(x) = c_1x_1 + \cdots + c_nx_n,$$

so

$$\frac{\partial F(x)}{\partial x_i} = c_i,$$

and again, we see the independence of the result on x_i.

5.1.2 Hopf Characteristic Functionals

To introduce Hopf functionals, we first require some notation. By $\langle h, \psi\rangle$ we mean

$$\langle h, \psi\rangle = \int_\Delta h(x)\psi(x)\,dx.$$

[Note that $\langle h, \psi\rangle \equiv \Psi(h)$.] Further, we let μ^ψ be a probabilistic measure on a space $C(\Delta)$ of functions ψ. By considering such a situation we are concentrating on the one in which a first choice gives a function, say ψ_1, while a second choice gives

ψ_2. We say that the functions ψ form a *random field* or that $C(\Delta)$ with μ^ψ forms a random field.

With these notions, we define the *Hopf characteristic functional* to be

$$\Phi(h) = \int_{C(\Delta)} e^{i\langle h,\psi\rangle}\, d\mu^\psi. \tag{5.3}$$

Note that the integration is over the space $C(\Delta)$. Furthermore, (5.3) can also be recognized as

$$\Phi(h) = \int_{C(\Delta)} e^{i\Psi(h)}\, d\mu^\psi.$$

Remark 5.2 In normal probability theory, if we have a random vector $(\xi_1, \cdots, \xi_n)$ and a vector $(\lambda_1, \cdots, \lambda_n)$, then the characteristic function is just

$$\begin{aligned} E\left(e^{i(\lambda_1\xi_1+\cdots+\lambda_n\xi_n)}\right) &= \int_\Omega e^{i(\lambda_1\xi_1+\cdots+\lambda_n\xi_n)}\, P_\xi(d\omega) \\ &= \int_\Omega e^{i\langle\lambda,\xi\rangle}\, dP_\xi \\ &= \phi(\lambda). \end{aligned}$$

Thus, replacing the discrete ξ_i by continuous functions makes the connection with the definition of the Hopf characteristic functional more evident:

$$E\left(e^{i\langle h,\psi\rangle}\right) = \int_{C(\Delta)} e^{i\langle h,\psi\rangle}\, d\mu^\psi \equiv \Phi(h).$$

Remark 5.3 The factor i in the exponential just serves to ensure that

$$e^{i\langle h,\psi\rangle}$$

is no greater than 1 since $|e^{ir}| \leq 1, r \in \mathbb{R}$.

To calculate the functional derivative of the Hopf characteristic functional, we must first calculate the increase of Φ between h and $\bar{h}$. In carrying out this computation we will need the fact that

$$e^{ix} - e^{iy} \simeq i(x-y)e^{i\tilde{x}}, \qquad \tilde{x} \underset{x\to y}{\longrightarrow} y.$$

Thus, we may write

$$\Phi(\bar{h}) - \Phi(h) = \int \left[e^{i\langle\bar{h},\psi\rangle} - e^{i\langle h,\psi\rangle}\right] d\mu^\psi$$

$$\simeq \int i\langle \bar{h} - h, \psi \rangle e^{\tilde{z}}\, d\mu^{\psi}.$$

Now the area between $\bar{h}$ and h is just

$$\Delta h = \int [\bar{h}(x) - h(x)]\, dx,$$

so

$$\frac{\Delta \Phi}{\Delta h} = \int i \left[\frac{\langle \bar{h} - h, \psi \rangle}{\Delta h} \right] e^{\tilde{z}}\, d\mu^{\psi}. \tag{5.4}$$

[N.B. Do not confuse the Δ here with the intervals introduced earlier.] Note that:

1. $\tilde{z} \to i\langle h, \psi \rangle$ as $\Delta h \to 0$;
2. By our considerations of the linear functional (5.1),

$$\frac{\langle \bar{h} - h, \psi \rangle}{\Delta h} \underset{\Delta h \to 0}{\longrightarrow} \psi(x_0).$$

Therefore, taking $\Delta h \to 0$ in (5.4) we have finally that the functional derivative of the Hopf characteristic functional (5.3) is given by

$$\frac{\delta \Phi(h)}{\delta h(x_0)} = i \int_{C(\Delta)} \psi(x_0) e^{i\langle h, \psi \rangle}\, d\mu^{\psi}. \tag{5.5}$$

Note from (5.5) that the functional differential of an integral behaves just like in the normal calculus since

$$\begin{aligned}
\frac{\delta \Phi(h)}{\delta h(x_0)} &= \frac{\delta}{\delta h(x_0)} \int_{C(\Delta)} e^{i\Psi(h)}\, d\mu^{\psi} \\
&= \int_{C(\Delta)} \frac{\delta}{\delta h(x_0)} e^{i\Psi(h)}\, d\mu^{\psi} \\
&= i \int_{C(\Delta)} \frac{\delta \Psi(h)}{\delta h(x_0)} e^{i\Psi(h)}\, d\mu^{\psi} \\
&= i \int_{C(\Delta)} \psi(x_0) e^{i\Psi(h)}\, d\mu^{\psi},
\end{aligned}$$

where we have used (5.2).

This last result, Eq. (5.5), is interesting in the following sense. To see why, assume that $h = 0$. Then we have

$$\frac{\delta \Phi(0)}{\delta h(x_0)} = i \int \psi(x_0)\, d\mu^{\psi}.$$

However,

$$E(\psi(x_0)) = \int \psi(x_0)\, d\mu^{\psi},$$

and, thus, for the Hopf characteristic functional we have

$$E(\psi(x_0)) = \frac{1}{i}\frac{\delta\Phi(0)}{\delta h(x_0)}. \tag{5.6}$$

It is interesting to pursue this procedure, for example by calculating

$$\frac{\delta^2\Phi(h)}{\delta h(x_1)\delta h(x_0)}.$$

By the same arguments as before, we have

$$\begin{aligned}
\Delta\left(\frac{\delta\Phi(h)}{\delta h(x_0)}\right) &= \frac{\delta\Phi(\bar{h})}{\delta h(x_0)} - \frac{\delta\Phi(h)}{\delta h(x_0)} \\
&= i\int \psi(x_0)\left[e^{i\langle\bar{h},\psi\rangle} - e^{i\langle h,\psi\rangle}\right] d\mu^{\psi} \\
&\simeq i^2 \int \psi(x_0)\langle\bar{h}-h,\psi\rangle e^{\tilde{z}}\, d\mu^{\psi}.
\end{aligned}$$

From this we have

$$\frac{\Delta\left(\dfrac{\delta\Phi(h)}{\delta h(x_0)}\right)}{\Delta h} \simeq i^2 \int \psi(x_0)\left[\frac{\langle\bar{h}-h,\psi\rangle}{\delta h}\right] e^{\tilde{z}}\, d\mu^{\psi}.$$

Taking the limit of this final expression as $\Delta h \to 0$ we obtain the second functional derivative of the Hopf functional which is given by

$$\frac{\delta^2\Phi(h)}{\delta h(x_1)\delta h(x_0)} = i^2 \int \psi(x_0)\psi(x_1)e^{i\langle h,\psi\rangle}\, d\mu^{\psi}.$$

Once again taking $h = 0$, we find

$$E(\psi(x_0)\psi(x_1)) = \frac{1}{i^2}\frac{\delta^2\Phi(0)}{\delta h(x_1)\delta h(x_0)} \tag{5.7}$$

so as a special case

$$E(\psi^2(x_0)) = \frac{1}{i^2}\frac{\delta^2\Phi(0)}{\delta h(x_0)^2}.$$

The consequences of Eqs. (5.6) and (5.7), and similar versions that can be easily derived using these elementary techniques are quite surprising. Namely, given the Hopf characteristic functional (5.3):

$$\Phi(h) = \int_{C(\Delta)} e^{i\langle h,\psi\rangle}\, d\mu^{\psi}$$

with respect to the probabilistic measure μ^{ψ} of a random field of functions ψ, we can calculate all the probabilistic properties (moments) of the random field by simply evaluating various functional derivatives of Φ at $h = 0$.

Remark 5.4 As a simple example of these ideas, consider all of the trajectories of British Airways flight 094 from Warsaw to London over a 10-year period. Each trajectory $\psi(x)$ gives the altitude ψ at location x, and $\psi(x)$ can be considered as a member of a random field. We can thus make some statements about this ensemble of trajectories. For example, if x_0 is a particular location between Warsaw and London, then the expected value of the altitude at x_0 is given by

$$E(\psi(x_0)) = \frac{1}{i}\frac{\delta\Phi(0)}{\delta h(x_0)},$$

the variance in the altitude at x_0 is given by

$$E(\psi^2(x_0)) - [E(\psi(x_0))]^2 = \frac{1}{i^2}\frac{\delta^2\Phi(0)}{\delta h(x_0)^2} - \left[\frac{1}{i}\frac{\delta\Phi(0)}{\delta h(x_0)}\right]^2,$$

and the correlation between the altitudes at points x_0 and x_1 is given by

$$E(\psi(x_0)\psi(x_1)) = \frac{1}{i^2}\frac{\delta^2\Phi(0)}{\delta h(x_1)\delta h(x_0)}.$$

Remark 5.5 Suppose we have a partial differential equation and an initial condition is chosen from a random field with measure μ^{ψ}. Then as time increases from zero, the measure evolves, and we can consider the flow of measures. This was exactly the point of view adopted by Foias and Prodi (1976) in their study of the Navier–Stokes hydrodynamic equations. However, Hopf (1952) had earlier realized the difficulties in carrying out this approach and proposed, instead, that one might consider the evolution of the characteristic functionals. When the Hopf point of view is adopted, we have to solve a functional differential equation involving $\Phi(h)$.

Remark 5.6 If we have a system of ordinary differential equations, then the Liouville equation describes the evolution of a density under the action of a flow generated by the system of ordinary differential equations. Alternately, we could form the Hopf functional differential equation for the evolution of $\Phi(h)$.

5.1.3 Hopf Characteristic Functionals and Partial Differential Equations

In this section we turn to a discussion of partial differential equations of the form

$$\frac{\partial u}{\partial t} = Lu \tag{5.8}$$

from the perspective of Hopf characteristic functionals. In (5.8) the operator L is assumed to be a linear combination of terms of the form

$$u, \quad \frac{\partial u}{\partial x_i}, \quad \frac{\partial^2 u}{\partial x_i \partial x_j}, \quad \ldots,$$

and products of these terms, with initial functions $u(0, x) = \psi(x)$.

For systems whose dynamics are governed by equations like (5.8), we can always derive a functional differential equation for $\Phi(u)$. Boundary value problems are rather delicate and difficult to treat using the Hopf method, but it is rather easy to treat initial value problems

$$\begin{aligned} \frac{\partial u}{\partial t} &= Lu \\ u(0, x) &= \psi(x). \end{aligned} \tag{5.9}$$

Here we assume that W is a space, $\psi \in W$, and that we have a probabilistic measure μ^ψ on W. How are we to view equations (5.8) within this context? Assume that we have a set of initial functions. Then Eq. (5.9) describes the evolution of functions such that a set or ensemble of initial functions evolves in time to a new ensemble. We assume that the evolution of these initial functions is such that the measure is preserved, i.e.

$$\mu^\psi(W_t) = \mu^\psi(W_0). \tag{5.10}$$

Having this probabilistic measure μ^ψ we couple (adjoin) it to the Hopf characteristic functional

$$\Phi_0(h) = \int e^{i\langle h, \psi \rangle} \, d\mu^\psi,$$

where the integration is over the space W but the notation has been suppressed. Likewise, to μ_t^ψ we adjoin

$$\Phi_t(h) = \int e^{i\langle h,\psi\rangle}\, d\mu_t^\psi.$$

How does this probabilistic measure evolve? As we have seen, knowing the Hopf characteristic functional gives a great deal of information concerning the random field and we will thus consider the evolution of measures via the evolution of $\Phi_t(h)$. Then, given $\Phi_t(h)$ we may return to a consideration of $\mu_t^\psi \equiv \mu^\psi(W_t)$.

Once again return to our basic system Eq. (5.9), so it is clear that the solution $u(t,x)$ depends on ψ which can be indicated by writing $u(t,\cdot)$, with the "·" indicating dependence on a whole function. The system (5.9) is equivalent to the operation of a transformation S_t operating on W (the space of initial functions ψ), $u(t,\cdot) = S_t(\cdot)$, and from Eq. (5.10) we may write

$$\mu^\psi(W_t) = \mu^\psi(W_0) = \mu^\psi(S_t^{-1}(W_t)).$$

Using this we have

$$\int e^{i\langle h,u(t,\cdot)\rangle}\, d\mu^\psi = \int e^{i\langle h,u(t,\cdot)\rangle}\, d\mu^\psi(S_t^{-1}(W_t)).$$

Changing the variables on the right-hand side gives

$$\int e^{i\langle h,u(t,\cdot)\rangle}\, d\mu^\psi = \int e^{i\langle h,S_t^{-1}(u(t,\cdot))\rangle}\, d\mu^\psi(W_t).$$

If we fix the time and set $u(t,\cdot) = S_t(\psi)$, this then becomes

$$\int e^{i\langle h,u(t,\cdot)\rangle}\, d\mu^\psi = \int e^{i\langle h,\psi\rangle}\, d\mu_t^\psi, \tag{5.11}$$

which is a very important and fundamental relationship. Equation (5.11) is entirely analogous to the "change of variables" formula (Lasota and Mackey 1994, Theorem 3.2.1). The only difference is that ψ now plays the role of x since we are integrating over functions.

Let us now fix the following notation. Write $\Phi_t(h) \equiv \Phi(t,h)$ so

$$\Phi(0,h) = \int e^{i\langle h,\psi\rangle}\, d\mu^\psi$$

and

$$\begin{aligned}\Phi(t,h) &= \int e^{i\langle h,\psi\rangle}\, d\mu_t^\psi \\ &= \int e^{i\langle h,u(t,\cdot)\rangle}\, d\mu^\psi.\end{aligned}$$

To study the evolution of the Hopf functional with respect to the system (5.9) we now differentiate $\Phi(t, h)$ with respect to t to give

$$\begin{aligned}
\frac{\partial \Phi}{\partial t} &= \frac{\partial}{\partial t} \int e^{i\langle h,u\rangle}\, d\mu^{\psi} \\
&= \int \frac{\partial}{\partial t} e^{i\langle h,u\rangle}\, d\mu^{\psi} \\
&= i \int \frac{\partial \langle h, u\rangle}{\partial t} e^{i\langle h,u\rangle}\, d\mu^{\psi} \\
&= i \int \left\langle h, \frac{\partial u}{\partial t} \right\rangle e^{i\langle h,u\rangle}\, d\mu^{\psi} \\
&= i \int \langle h, Lu\rangle e^{i\langle h,u\rangle}\, d\mu^{\psi} \\
&= i \int \left\{ \int h(x) Lu(t, x)\, dx \right\} e^{i\langle h,u\rangle}\, d\mu^{\psi} \\
&= i \int h(x) \left[\int Lu e^{i\langle h,u\rangle}\, d\mu^{\psi} \right] dx.
\end{aligned}$$

Therefore, our final formula becomes

$$\frac{\partial \Phi}{\partial t} = i \int h(x) \left[\int Lu e^{i\langle h,u\rangle}\, d\mu^{\psi} \right] dx, \tag{5.12}$$

which we will illustrate through a series of examples.

Example 5.2 Let us consider the partial differential equation

$$\frac{\partial u}{\partial t} = u. \tag{5.13}$$

From Eq. (5.12), since $Lu = u$, we have

$$\frac{\partial \Phi}{\partial t} = i \int h(x) \left[\int u e^{i\langle h,u\rangle}\, d\mu^{\psi} \right] dx.$$

However,

$$\frac{\delta \Phi(t, h)}{\delta h(x)} = i \int u e^{i\langle h,u\rangle}\, d\mu^{\psi},$$

so the functional differential equation for Φ corresponding to (5.13) is simply

$$\frac{\partial \Phi}{\partial t} = \int h(x) \frac{\delta \Phi}{\delta h(x)} \, dx. \tag{5.14}$$

The solution of Eq. (5.14) is

$$\Phi(t, h) = \Phi_0(e^t h), \tag{5.15}$$

where $\Phi(0, h) = \Phi_0(h)$ is the characteristic functional of the initial measures.

Before showing how we arrived at the solution, let us just go through the exercise of verifying that it is indeed the solution. First, recall that

$$\Phi(\bar{g}) - \Phi(g) = \int \frac{\delta \Phi(g)}{\delta h(x)} [\bar{g}(x) - g(x)] \, dx + O\left(\int |\bar{g}(x) - g(x)| \, dx \right).$$

Now we have, setting $g = e^t h$ and using (5.15),

$$\begin{aligned}
\frac{\partial \Phi(t, h)}{\partial t} &= \lim_{\bar{t} \to t} \frac{\Phi_0(e^{\bar{t}} h) - \Phi_0(e^t h)}{\bar{t} - t} \\
&= \lim_{\bar{t} \to t} \left\{ \int [e^{\bar{t}} - e^t] h(x) \frac{\delta \Phi_0(f)}{\delta h(x)} \bigg|_{f = e^t h} dx + O\left(\int |e^{\bar{t}} h - e^t h| \, dx \right) \right\} \\
&\simeq \lim_{\bar{t} \to t} \left\{ \int [e^{\bar{t}} - e^t] h(x) \frac{\delta \Phi_0(f)}{\delta h(x)} \bigg|_{f = e^t h} dx + O\left([e^{\bar{t}} - e^t] \int h(x) \, dx \right) \right\} \\
&= e^t \int h(x) \frac{\delta \Phi_0(f)}{\delta h(x)} \bigg|_{f = e^t h} dx.
\end{aligned} \tag{5.16}$$

As an aside, note that if we have a general functional $\Phi(\lambda h)$, then

$$\begin{aligned}
\frac{\delta [\Phi(\lambda h)]}{\delta h(x)} &= \lim_{\bar{h} \to h} \frac{\Phi(\lambda \bar{h}) - \Phi(\lambda h)}{\bar{h} - h} \\
&= \lambda \lim_{\bar{h} \to h} \frac{\Phi(\lambda \bar{h}) - \Phi(\lambda h)}{\lambda \bar{h} - \lambda h} \\
&= \lambda \frac{\delta \Phi}{\delta h(x)} (\lambda h) = \lambda \frac{\delta \Phi(f)}{\delta h(x)} \bigg|_{f = \lambda h}.
\end{aligned}$$

Thus, taking $\lambda = e^t$ we may write

$$\frac{\delta [\Phi_0(e^t h)]}{\delta h(x)} = e^t \frac{\delta \Phi_0(f)}{\delta h(x)} \bigg|_{f = e^t h},$$

and, therefore, (5.16) becomes

$$\frac{\partial \Phi(t,h)}{\partial t} = \int h(x) \frac{\delta[\Phi_0(e^t h)]}{\delta h(x)}\, dx$$
$$= \int h(x) \frac{\delta \Phi(t,h)}{\delta h(x)}\, dx,$$

thereby demonstrating that (5.15) is indeed the solution of the functional differential equation (5.14).

Now that we have verified the solution (5.15), let us consider the problem of how one could have obtained (5.15) without making a series of random guesses. This turns out to be quite straightforward since there is a general technique for solving Hopf functional differential equations corresponding to linear evolution equations.

Specifically, again consider the evolution equation

$$\frac{\partial u}{\partial t} = Lu, \tag{5.17}$$

with the initial condition $u(0,\cdot) = \psi$, where Lu is linear, and the corresponding Hopf functional differential equation is

$$\frac{\partial \Phi}{\partial t} = i \int h(x) \left[\int Lu e^{i\langle h,u\rangle}\, d\mu^{\psi} \right] dx \tag{5.18}$$

with a characteristic functional $\Phi(0,h) = \Phi_0(h)$ of the initial measure. If the solution of Eq. (5.17) is written as

$$u(t,\cdot) = T_t \psi \tag{5.19}$$

(T_t is now a linear operator), then the solution of the corresponding Hopf equation is simply

$$\Phi(t,h) = \Phi_0(T_t^* h), \tag{5.20}$$

where T_t^* is the operator adjoint to T_t, i.e., the operator satisfying

$$\langle h, T\psi \rangle = \langle T^* h, \psi \rangle.$$

To show that (5.20) is indeed the solution to the Hopf equation (5.18) is quite straightforward. Thus, if we start from

$$\Phi_0(h) = \int e^{i\langle h,\psi\rangle}\, d\mu^{\psi},$$

then

$$\Phi(t,h) = \int e^{i\langle h, u(t,\cdot)\rangle}\, d\mu^{\psi}. \tag{5.21}$$

From Eq. (5.19), $u(t,\cdot) = T_t\psi$, so Eq. (5.21) becomes

$$\begin{aligned}\Phi(t,h) &= \int e^{i\langle h, T_t h\rangle}\, d\mu^{\psi} \\ &= \int e^{i\langle T_t^* h, \psi\rangle}\, d\mu^{\psi} \\ &= \Phi_0(T_t^* h),\end{aligned}$$

thus demonstrating equation (5.20).

With this discussion under our belt, we can now consider a second example.

Example 5.3 Consider the initial value problem

$$\frac{\partial u}{\partial t} = a(t,x)\frac{\partial u}{\partial x} \tag{5.22}$$

with $u(0,\cdot) = \psi(\cdot)$. From the general equation (5.12), we have directly that

$$\begin{aligned}\frac{\partial \Phi}{\partial t} &= i\int h(x)\left[\int a(t,x)\frac{\partial u}{\partial x} e^{i\langle h,u\rangle}\, d\mu^{\psi}\right] dx \\ &= \int h(x)a(t,x)\frac{\partial}{\partial x}\left[i\int u e^{i\langle h,u\rangle}\, d\mu^{\psi}\right] dx\end{aligned}$$

so, using Eq. (5.5),

$$\frac{\partial \Phi}{\partial t} = \int h(x)a(t,x)\frac{\partial}{\partial x}\frac{\delta \Phi}{\delta h(x)} dx$$

is the functional differential equation for Φ corresponding to (5.22).

It is equally easy to show that the slightly more general initial value problem

$$\frac{\partial u}{\partial t} = a(t,x)\frac{\partial u}{\partial x} + b(t,x)u \tag{5.23}$$

has a corresponding Hopf equation

$$\frac{\partial \Phi}{\partial t} = \int h(x)\left[a(t,x)\frac{\partial}{\partial x}\frac{\delta \Phi}{\delta h(x)} + b(t,x)\frac{\delta \Phi}{\delta h(x)}\right] dx.$$

Example 5.4 If we take the particular case of $a(t,x) = -x$ and $b(t,x) = \lambda$, then (5.23) becomes

$$\frac{\partial u}{\partial t} = -x\frac{\partial u}{\partial x} + \lambda u, \tag{5.24}$$

which is a linearized version of an equation that has been applied to the problem of describing the simultaneous proliferation and maturation of a population of cells [see Lasota and Mackey (1994, Example 11.1.1, pp. 297–302), where it is shown that the solutions of (5.24) are exact with respect to the Wiener measure].

The solution of (5.24), given an initial condition $u(0,\cdot) = \psi(\cdot)$, is simply

$$u(t,x) = e^{\lambda t}\psi(e^{-t}x), \tag{5.25}$$

which may be obtained using the method of characteristics. Thus, writing (5.25) as

$$u(t,\cdot) = T_t\psi(\cdot),$$

where

$$(T_t\psi)(x) = e^{\lambda t}\psi(e^{-t}x),$$

we also have

$$\langle h, T_t\psi\rangle = \int h(x)e^{\lambda t}\psi(e^{-t}x)\,dx.$$

Make the change of variables $y = e^{-t}x$ so $dx = e^t dy$ and thus

$$\langle h, T_t\psi\rangle = \int e^{(\lambda+1)t}h(e^t y)\psi(y)\,dy.$$

Thus, it is straightforward to see that

$$(T_t^* h)(x) = e^{(\lambda+1)t}h(e^t x),$$

and, therefore, the Hopf equation

$$\frac{\partial \Phi}{\partial t} = \int h(x)\left[-x\frac{\partial}{\partial x}\frac{\delta\Phi}{\delta h(x)} + \lambda\frac{\delta\Phi}{\delta h(x)}\right]dx$$

corresponding to (5.24) has the solution

$$\Phi(t,h) = \Phi_0(e^{(\lambda+1)t}h(e^t\cdot)),$$

where Φ_0 is the characteristic functional of the initial measure.

Example 5.5 Consider the system of ordinary differential equations

$$\frac{dx_k}{dt} = F_k(x), \qquad k = 1, \cdots, d \tag{5.26}$$

and the corresponding Liouville equation for the evolution of the density $f(x, t)$ under the action of the flow generated by (5.26):

$$\frac{\partial f}{\partial t} = -\sum_{k=1}^{d} \frac{\partial [F_k f]}{\partial x_k}. \tag{5.27}$$

Rewriting (5.27) as

$$\frac{\partial f}{\partial t} = -f \sum_{k=1}^{d} \frac{\partial F_k}{\partial x_k} - \sum_{k=1}^{d} F_k \frac{\partial f}{\partial x_k},$$

and identifying L in an obvious manner, we have from (5.12) that

$$\frac{\partial \Phi}{\partial t} = -i \int h(x) \left[\int \sum_{k=1}^{d} \frac{\partial F_k}{\partial x_k} u e^{i\langle h,u \rangle} d\mu^{\psi} - \sum_{k=1}^{d} F_k \frac{\partial}{\partial x_k} \int u e^{i\langle h,u \rangle} d\mu^{\psi} \right] dx.$$

Then, using (5.5) and some simple manipulations it is almost immediate that the differential equation for the Hopf functional Φ is given by

$$\frac{\partial \Phi}{\partial t} = -\int h(x) \left[\sum_{k=1}^{d} \frac{\partial}{\partial x_k} \left(F_k \frac{\delta \Phi}{\delta h(x)} \right) \right] dx.$$

5.2 Characteristic Functionals for Delay Equations

This section is taken in its entirety from Losson (1991) which was later published in Losson and Mackey (1992).

We consider differential delay equations of the form

$$\frac{du}{ds} = -\alpha u(s) + S(u(s-1)) \text{ for } 1 < s, \tag{5.28}$$

in which the delay τ is taken to be 1 without loss of generality, with the initial function

$$u(s) = v(s) \text{ if } s \in [0, 1]. \tag{5.29}$$

From now on we write Eqs. (5.28) and (5.29) as the combined system

$$\begin{cases} u(s) = v(s) & \text{for } s \in [0, 1], \\ \dfrac{du(s)}{ds} = -\alpha u(s) + S(v(s-1)) & \text{for } 1 < s < 2 \end{cases} \tag{5.30}$$

and denote by S_t the corresponding semi-dynamical system $S_t\colon C([0,1]) \longmapsto C([0,1])$ given by

$$S_t v(x) = u_v(x+t), \tag{5.31}$$

where u_v denotes the solution of (5.30) corresponding to the initial function v. Equation (5.31) defines a *semi-dynamical system* because a differential delay equation is *noninvertible*, i.e., it cannot be run unambiguously forward and backwards in time.

From Eq. (5.31), the system (5.30) is equivalent to considering

$$\frac{\partial}{\partial t} S_t v(x) = \begin{cases} \frac{\partial}{\partial t} v(x+t) & \text{for } x \in [0, 1-t], \\ -\alpha u(x+t) + S(v(x+t-1)) & \text{for } x \in (1-t, 1]. \end{cases} \tag{5.32}$$

Thus, we consider a segment of a solution of (5.30) defined on an interval $I_t = [t, t+1]$, as t increases (continuously) [i.e. the differential delay equation (5.30) operates on a *buffer* of length 1, "sliding" it along the time axis]. Equation (5.32) states that the contents of this buffer are the initial condition v when the argument $(x+t)$ is less than 1, and the solution u of the equation otherwise.

We next introduce the characteristic functional $\mathcal{Z}_t$ of a family of probability measures evolving from an initial measure. We define the characteristic functional $\mathcal{Z}_t$ for (5.32) by

$$\mathcal{Z}_t[J_1, J_2] = \int_C \exp\left[i \int_0^1 J_1(x) u_v(x+t)\, dx + i \int_0^1 J_2(x) v(x)\, dx \right] d\mu_0(\mathcal{T}_t^{-1}(v, u_v)). \tag{5.33}$$

The source functions J_1 and J_2 are elements of $C([0,1])$ and the measure of integration is the initial measure μ_0 (describing the initial distribution of functions) composed with $\mathcal{T}_t^{-1}(v, u_v)$ where $\mathcal{T}_t(v) : C([0,1]) \longmapsto C \times C$ is defined by

$$\mathcal{T}_t(v) = (v, u_v). \tag{5.34}$$

For simplicity, we will use the notation $\mu_0(\mathcal{T}_t^{-1}(v, u_v)) \equiv \mathcal{W}[v, S_t(v)]$, so (5.33) becomes

$$\mathcal{Z}_t[J_1, J_2] = \int_C \exp\left[i \int_0^1 J_1(x) u_v(x+t)\, dx + i \int_0^1 J_2(x) v(x)\, dx \right] d\mathcal{W}[v, S_t(v)]. \tag{5.35}$$

When no confusion is possible, we write $\mathcal{W}_t$ for $\mathcal{W}[v, S_t(v)]$.

If f and g are two functions defined on an interval I, we denote their scalar product by

$$\langle f, g \rangle \equiv \int_I f(x) g(x)\, dx.$$

To simplify the notation, we also write

$$\Upsilon[J_1, J_2; v] = \exp\left[i \langle J_1(x), u_v(x+t) \rangle + i \langle J_2(x), v(x) \rangle \right]. \tag{5.36}$$

Υ is used from now on to denote the function of J_1, J_2 and v defined in (5.36). We begin by noting the following relations:

$$\frac{\delta^n \mathcal{Z}_t}{\delta J_1^n(\xi)} = i^n \left\langle \Upsilon u_v^n(\xi + t) \right\rangle, \tag{5.37}$$

$$\frac{\delta^n \mathcal{Z}_t}{\delta J_2^n(\xi)} = i^n \left\langle \Upsilon v^n(\xi) \right\rangle, \tag{5.38}$$

where it is understood that

$$\left\langle \left(\vdots \right) \right\rangle = \int_C \left(\vdots \right) \, d\mathcal{W}[v, S_t(v)].$$

Note that if μ_0 is the probability measure on the space of all initial functions v, and A is any subset of $C([0, 1])$, then

$$\mu_t(A) \equiv \mu_0(S_t^{-1}(A)).$$

In other words, the probability that a randomly chosen function belongs to A at time t equals the probability that the counterimage of that function (under the action of S_t) belonged to the counterimage of the set A. This defines the family of measures characterized by the solutions $\mathcal{Z}_t$ of a functional differential equation which is the Fourier transform of the infinite-dimensional version of the Kramers–Moyal expansion (Risken 1984). The derivation of such an equation for a differential delay equation was first considered by Capiński (1991). If the semi-flow S_t is measure-preserving with respect to μ_0, then $\mu_0(A) \equiv \mu_0(S_t^{-1}(A))$. In this case, we alternately say that the measure μ_0 is invariant with respect to S_t.

We are now able to derive an evolution equation for the characteristic functional.

5.2.1 A Functional Differential Equation for $\mathcal{Z}_t$

Time differentiation of the characteristic functional $\mathcal{Z}_t$ defined in (5.35) yields, in conjunction with (5.32),

$$\begin{aligned}\frac{\partial \mathcal{Z}_t}{\partial t} &= i\left\langle \Upsilon \int_0^1 J_1(x)\frac{\partial u_v(x+t)}{\partial t}\,dx\right\rangle \\ &= i\left\langle \Upsilon \int_0^1 J_1(x)\frac{\partial u_v(x+t)}{\partial x}\,dx\right\rangle \\ &= i\left\langle \Upsilon \int_0^{1-t} J_1(x)\frac{\partial v(x+t)}{\partial x}\,dx - \alpha\Upsilon\int_{1-t}^1 J_1(x)u(x+t)\,dx\right\rangle \\ &\quad +i\left\langle \Upsilon \int_{1-t}^1 J_1(x)S(v(x+t-1))\,dx\right\rangle .\end{aligned}$$

Therefore, from Eq. (5.37) and the definition (5.32), we obtain

$$\begin{aligned}\frac{\partial \mathcal{Z}_t}{\partial t} &= \int_0^{1-t} J_1(x)\frac{\partial}{\partial x}\left(\frac{\delta \mathcal{Z}_t}{\delta J_1(x)}\right)dx - \alpha\int_{1-t}^1 J_1(x)\frac{\delta \mathcal{Z}_t}{\delta J_1(x)}\,dx \\ &\quad +i\left\langle \Upsilon \int_{1-t}^1 J_1(x)S(v(x+t-1))\,dx\right\rangle . \end{aligned} \tag{5.39}$$

Equation (5.39) is related to the Hopf functional differential equation for the evolution of the characteristic functional $\mathcal{Z}_t$, and contains all the statistical information describing the evolution of a density of initial functions under the action of the differential delay system (5.28) and (5.29). An equation similar to (5.39) was first obtained by Capiński (1991) for a differential delay equation with a quadratic nonlinearity (see Example 5.6 below).

In order to derive the Hopf equation per se, we restrict our attention to situations where the feedback function S in the differential delay equation (5.28) is a polynomial

$$S(v) = \sum_{k=1}^n a_k v^k. \tag{5.40}$$

With nonlinearity (5.40), Eq. (5.39) becomes, with identity (5.38),

$$\begin{aligned}\frac{\partial \mathcal{Z}_t}{\partial t} &= \int_0^{1-t} J_1(x)\frac{\partial}{\partial x}\left(\frac{\delta \mathcal{Z}_t}{\delta J_1(x)}\right)dx - \alpha\int_{1-t}^1 J_1(x)\frac{\delta \mathcal{Z}_t}{\delta J_1(x)}\,dx \\ &\quad +\sum_{k=1}^n i^{(1-k)}\,a_k\int_{1-t}^1 J_1(x)\frac{\delta^k \mathcal{Z}_t}{\delta J_2^k(x+t-1)}\,dx. \end{aligned} \tag{5.41}$$

Analytically solving the Hopf equation (5.41) is not possible at present, though a correct method of solution should make use of integration with respect to measures defined on the space C. Presently, the theory of such integrals only allows their consistent utilization in solving functional differential equations when the measure of integration is the Wiener measure (Sobczyk 1984).

Remark 5.7 Note that this is the point where we stumble in our attempts to use the Hopf functional approach, and this was one of the stumbling blocks mentioned in Chap. 4.

Before proceeding to treat the Hopf equation in a perturbative manner, we illustrate its specific form for a simple nonlinear delay equation.

Example 5.6 (Capiński (1991)) The differential delay equation

$$\frac{du}{ds} = -\alpha u(s) + ru(s-1)[1-u(s-1)], \tag{5.42}$$

can be considered as a continuous analogue of the discrete time quadratic map

$$u_{n+1} = ru_n(1-u_n) \tag{5.43}$$

because Eq. (5.42) is the *singular perturbation* of the quadratic map (5.43) as defined in Ivanov and Šarkovskiĭ (1991). The characteristic functional is defined by (5.35), and the functional differential equation corresponding to (5.41) was shown by Capiński (1991) to be

$$\begin{aligned}
\frac{\partial \mathcal{Z}_t}{\partial t} &= \int_0^{1-t} J_1(x)\frac{\partial}{\partial x}\left(\frac{\delta \mathcal{Z}_t}{\delta J_1(x)}\right) dx - \alpha \int_{1-t}^{1} J_1(x)\frac{\delta \mathcal{Z}_t}{\delta J_1(x)}\, dx \\
&+r \int_{1-t}^{1} J_1(x)\frac{\delta \mathcal{Z}_t}{\delta J_2(x+t-1)}\, dx \\
&-ri^{-1} \int_{1-t}^{1} J_1(x)\frac{\delta^2 \mathcal{Z}_t}{\delta J_2^2(x+t-1)}\, dx.
\end{aligned} \tag{5.44}$$

In spite of the fact that we cannot solve the Hopf equation analytically, relatively mild assumptions allow one to gain significant insight into the dynamics of $\mathcal{Z}_t$. More precisely, if $\mathcal{Z}_t$ is analytic, we can expand it in a power series and treat the Hopf equation in a perturbative manner. We follow this approach in the next section.

5.2.2 The Moments of the Measure $\mathcal{W}_t$

The statistical properties of the random field of functions v and u are described by an infinite hierarchy of moments of the measure $\mathcal{W}_t$. For fixed t, the average value of the contents of the buffer defined on $I_t = [t, t+1]$ (i.e., v on $[t, 1]$ and u_v on $(1, 1+t]$), which is just the first order moment of the measure $\mathcal{W}_t$, is

$$M_v^1(t, x) \equiv \int_C v(x+t)\, d\mu_0(v) \qquad \text{for } x \in [0, 1-t], \tag{5.45}$$

$$M_u^1(t, x) \equiv \int_C u_v(x+t)\, d\mu_0(v) \qquad \text{for } x \in (1-t, 1]. \tag{5.46}$$

These two equations can be written as one relation:

$$M^1(t, x) \equiv \int_C u_v(x+t)\, d\mathcal{W}_t \qquad \text{for } x \in [0, 1].$$

The definition of the second order moment (or covariance function) $M^2(t, x, y)$ is, with the same notation,

$$M^2(t, x, y)$$

$$= \begin{cases} \int_C v(x+t)v(y+t)d\mu_0(v) \equiv M_{vv}^2(t, x, y) : x, y \in [0, 1-t] \times [0, 1-t], \\ \int_C u_v(x+t)v(y+t)d\mu_0(v) \equiv M_{uv}^2(t, x, y) : x, y \in (1-t, 1] \times [0, 1-t], \\ \int_C v(x+t)u_v(y+t)d\mu_0(v) \equiv M_{vu}^2(t, x, y) : x, y \in [0, 1-t] \times (1-t, 1], \\ \int_C u_v(x+t)u_v(y+t)d\mu_0(v) \equiv M_{uu}^2(t, x, y) : x, y \in (1-t, 1] \times (1-t, 1]. \end{cases}$$

The subscripts of the various components of M^2 refer to the segments of the solution whose correlation is given by the particular component. For example, M_{uv}^2 describes the correlation between u and v segments of the solution as is illustrated in Fig. 5.1. Remember that the initial function is defined on an interval [0, 1], so to complete the description of the statistical dependence of the solution u_v on the initial function it is necessary to introduce the functions M_{ou}^2. M_o^1 is the first order moment of measure μ_0, M_{oo}^2 is the second order moment of μ_0, etc.

The moments of the measure $\mathcal{W}_t$ are also given by the power series expansion of the characteristic functional $\mathcal{Z}_t$ as we next discuss.

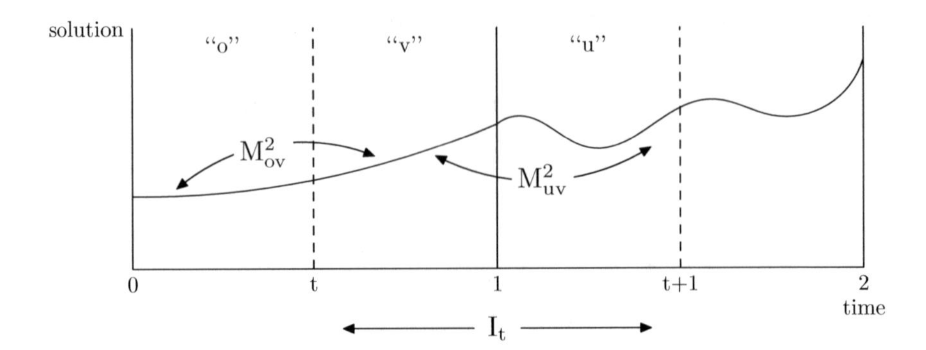

Fig. 5.1 A differential delay equation transforms a function defined on [0, 1] into a function defined on I_t. Illustration of the "o," "v," and "u" segments of the solution. Reproduced from Losson and Mackey (1992) with permission

5.2.3 Taylor Series Expansion of the Functional $\mathcal{Z}_t$

The expression for the series expansion of a functional can be understood with the following argument. Let

$$F(y_1, \cdots, y_k) = F(\mathbf{y})$$

be a function of k variables. The power series expansion of F is

$$F(\mathbf{y}) = \sum_{n=0}^{\infty} \sum_{i_1=0}^{k} \cdots \sum_{i_n=0}^{k} \frac{1}{n!} \mathcal{E}_n(i_1, \cdots, i_n)(y_1, \cdots, y_n),$$

where

$$\mathcal{E}_n(i_1, \cdots, i_n) = \left. \frac{\partial^n F(\mathbf{y})}{\partial y_1 \cdots \partial y_n} \right|_{\mathbf{y}=0}.$$

Passing to a continuum in the following sense

$$\begin{aligned} i &\longrightarrow x_i, \\ y_i(i = 1, \cdots, k) &\longrightarrow y(x), \\ -\infty < \; &x \; < \infty, \\ \sum_i &\longrightarrow \int_{\mathbb{R}} dx, \end{aligned}$$

we obtain the corresponding series expansion of a *functional* Ψ:

$$\Psi[y] = \sum_{n=0}^{\infty} \int_{\mathbb{R}^n} dx_1 \cdots dx_n \mathcal{E}_n(x_1, \cdots, x_n) y(x_1) \cdots y(x_n),$$

where

$$\mathcal{E}_n(x_1, \cdots, x_n) = \frac{1}{n!} \frac{\delta^n \Psi[y]}{\delta y(x_1) \cdots \delta y(x_n)} \bigg|_{y=0}.$$

$\Psi[y]$ is called the *characteristic functional* of the functions $\mathcal{E}_n$.

With these conventions, the expansion of the characteristic functional (5.35) is

$$\mathcal{Z}_t[J_1, J_2] = \sum_{p=0}^{\infty} \sum_{q=0}^{p} \int_0^1 \cdots \int_0^1 \mathcal{E}_{pq}(t, x_1, \cdots, x_p) \times \left(\prod_{j=1}^{q} J_1(x_j)\, dx_j \right) \left(\prod_{j=q+1}^{p} J_2(x_j)\, dx_j \right). \tag{5.47}$$

The kernels $\mathcal{E}_{pq}$ in the expansion are proportional to the moment functions of the measure $\mathcal{W}[v, S_t v]$. From Eqs. (5.37) and (5.38) they are given by

$$\begin{aligned} \mathcal{E}_{pq}(t, x_1, \cdots, x_p) &= \frac{1}{p!} \frac{\delta^p \mathcal{Z}_t}{\delta J_1^q \, \delta J_2^{p-q}} \\ &= \frac{i^p}{p!} \langle u_v(x_1) \cdots u_v(x_q) v(x_{q+1}) \cdots v(x_p) \rangle \\ &= \frac{i^p}{p!} M^p_{u^q v^{(p-q)}}(t, x_1, \cdots, x_p), \end{aligned} \tag{5.48}$$

where from now on we use the notation $M^p_{u^q v^{(p-q)}}(t, x_1, \cdots, x_p) = M^p_{u^q v^{(p-q)}}(t, \mathbf{x})$. Equation (5.47) is the infinite-dimensional generalization of the well-known expansion of a characteristic function in terms of the corresponding probability moments (or their Legendre transforms, the cumulants).

5.2.4 *Partial Differential Equations for the Moments*

The evolution equation of the kth moment is given by substituting the moment in question into (5.41) and then using formula (5.47) to the appropriate order.

Consider the first order moments of the measure $\mathcal{W}_t$. If we substitute the definitions (5.48) and the expansion (5.47) into Eq. (5.41), we obtain a partial differential equation for the moment $M^1(t, x)$:

$$\frac{\partial}{\partial t} M_v^1(t,x) = \frac{\partial}{\partial x} M_v^1(t,x) \quad \text{for } x \in [0, 1-t],$$

$$\frac{\partial}{\partial t} M_u^1(t,x) = -\alpha M_u^1(t,x) + \sum_{k=1}^{n} a_k M_{o^k}^k(x+t-1, \overset{k}{\ldots}, x+t-1)$$
$$\text{for } x \in (1-t, 1]. \tag{5.49}$$

Equation (5.49) is simply the Hopf equation (5.41) for the first order moments. In (5.49) the k arguments of $M_{o^k}^k$ indicate that it is the k-point autocorrelation function of the initial function distribution described by μ_0. Moments whose label does not contain u are moments of the initial measure.

The second order moment functions $M^2(t,x)$ are given by the solutions of the four equations:

$$\frac{\partial}{\partial t} M_{vv}^2(t,x,y) = \frac{\partial}{\partial x} M_{vv}^2(t,x,y) + \frac{\partial}{\partial y} M_{vv}^2(t,x,y)$$
$$\text{for } (x,y) \in [0, 1-t) \times [0, 1-t), \tag{5.50}$$

$$\frac{\partial}{\partial t} M_{uv}^2(t,x,y) = \frac{\partial}{\partial y} M_{uv}^2(t,x,y) - \alpha M_{uv}^2(t,x,y)$$
$$+ \sum_{k=2}^{n} a_k M_{o^{(k-1)}v}^k(t, x+t-1, \overset{(k-1)}{\ldots}, x+t-1, y)$$
$$\text{for } (x,y) \in (1-t] \times [0, 1-t], \tag{5.51}$$

$$\frac{\partial}{\partial t} M_{vu}^2(t,x,y) = \frac{\partial}{\partial x} M_{vu}^2(t,x,y) - \alpha M_{vu}^2(t,x,y)$$
$$+ \sum_{k=2}^{n} a_k M_{vo^{(k-1)}}^k(t, x, y+t-1, \overset{(k-1)}{\ldots}, y+t-1)$$
$$\text{for } (x,y) \in [0, 1-t] \times (1-t, 1], \tag{5.52}$$

$$\frac{\partial}{\partial t} M_{uu}^2(t,x,y) = -2\alpha M_{uu}^2(t,x,y)$$
$$+ \sum_{k=1}^{n} a_k \{ M_{o^{(k-1)}u}^k(t, x+t-1, \overset{(k-1)}{\ldots}, x+t-1, y)$$
$$+ M_{uo^{(k-1)}}^k(t, x, y+t-1, \overset{(k-1)}{\ldots}, y+t-1) \},$$
$$\text{for } (x,y) \in (1-t, 1] \times (1-t, 1]. \tag{5.53}$$

The functions M_{ou}^2 and M_{oou}^2 are given by

$$\frac{\partial}{\partial t} M_{ou}^2(t,x,y) = -\alpha M_{ou}^2$$

$$+\sum_{k=2}^{n} a_k M^k_{o^k}(x, y+t-1, .\overset{k}{.}., y+t-1), \tag{5.54}$$

$$\frac{\partial}{\partial t} M^3_{oou}(t, x, y, z) = -\alpha M^3_{oou}(t, x, y, z) +$$

$$+\sum_{k=3}^{n} a_k M^k_{o^k}(x, y, z+t-1, .\overset{k}{.}., z+t-1), \tag{5.55}$$

and similar equations give the moments $M^k_{o^{(k-1)}u}$.

A pattern clearly emerges from the preceding analysis: The moment $M^p(t, \mathbf{x})$ is given by 2^p partial differential equations of the same form as (5.50) through (5.53) since $M^p(t, \mathbf{x})$ is a function of p variables, each of which can belong to one of two possible intervals ($[0, 1-t]$ or $(1-t, 1]$). The first of these equations (when all the x_k's belong to $[0, 1-t]$) is

$$\frac{\partial}{\partial t} M^p_{v^p}(t, \mathbf{x}) = \sum_{j=1}^{p} \frac{\partial}{\partial x_j} M^p_{v^p}(t, \mathbf{x}). \tag{5.56}$$

We call the equations which give the moments of the form $M^p_{v^l u^{(p-l)}}$ *mixed* equations because they yield functions which correlate mixed u and v segments of the solution. For the moment of order p, there are $(2^p - 2)$ mixed equations and 2 *pure* equations. The pure equations give $M^p_{v^p}$ and $M^p_{u^p}$, the p-point autocorrelation functions of the v and u segments of the solution.

If $x_j \in [0, 1-t]$ for $j = 1, \cdots, l$ and $x_j \in (1-t, 1]$ for $j = l+1, \cdots, p$, then when the forcing term S of Eq. (5.28) is the polynomial (5.40), the generic form of the mixed equation for $M_{v^l u^{(p-l)}}$ is

$$\frac{\partial}{\partial t} M^p_{v^l\, u^{(p-l)}}(t, \mathbf{x}) = \sum_{i=1}^{l} \frac{\partial}{\partial x_i} M^p_{v^l\, u^{(p-l)}}(t, \mathbf{x}) - \alpha(p-l)\, M^p_{v^l\, u^{(p-l)}}(t, \mathbf{x})$$

$$+\sum_{j=0}^{n-1} a_j \left\{ M^{(p+j)}_{v^l\, o^j\, u^{(p-l)}}(t, \mathbf{x}) + M^{(p+j)}_{v^l\, u^{(p-l)}\, o^j}(t, \mathbf{x}) \right\}.$$

Once again, this equation is one representative of the $(2^p - 2)$ mixed equations to be solved to obtain the moment of order p. Deriving these equations is tedious, but the task is greatly simplified by the similarity existing between the systems of equations for moments of different orders.

Before proceeding, we illustrate the ideas presented above and derive the partial differential equations analogous to (5.49) and (5.50) through (5.53) for the nonlinear differential delay equation (5.42) considered in Example 5.6.

Example 5.7 When the differential delay equation is

$$\frac{du}{ds} = -\alpha u(s) + ru(s-1) - ru^2(s-1), \tag{5.57}$$

the first order moment equations are given by

$$\frac{\partial M_v^1(t,x)}{\partial t} = \frac{\partial M_v^1(t,x)}{\partial x}, \tag{5.58}$$

$$\frac{\partial M_u^1(t,x)}{\partial t} = -\alpha M_u^1(t,x) + rM_o^1(x+t-1) - rM_{oo}^1(x+t-1, x+t-1). \tag{5.59}$$

The four evolution equations for the second order moments are

$$\frac{\partial M_{vv}^2(t,x,y)}{\partial t} = \frac{\partial M_{vv}^2(t,x,y)}{\partial x} + \frac{\partial M_{vv}^2(t,x,y)}{\partial y}, \quad \text{for} \quad x, y \in [0, 1-t] \tag{5.60}$$

$$\begin{aligned} \frac{\partial M_{vu}^2(t,x,y)}{\partial t} &= \frac{\partial M_{vu}^2(t,x,y)}{\partial x} - \alpha M_{vu}^2(t,x,y) \\ &+ rM_{vo}^2(t,x,y+t-1) \\ &- rM_{voo}^3(t,x,y+t-1,y+t-1), \\ &\text{for} \quad x \in [0,1-t], \ y \in (1-t, 1] \end{aligned} \tag{5.61}$$

$$\begin{aligned} \frac{\partial M_{uv}^2(t,x,y)}{\partial t} &= \frac{\partial M_{uv}^2(t,x,y)}{\partial y} - \alpha M_{uv}^2(t,x,y) \\ &+ rM_{ov}^2(t,x+t-1,y) \\ &- rM_{oov}^3(t,x+t-1,x+t-1,y), \\ &\text{for} \quad x \in (1-t, 1], \ y \in [0,1-t] \end{aligned} \tag{5.62}$$

$$\begin{aligned} \frac{\partial M_{uu}^2(t,x,y)}{\partial t} &= -2\alpha M_{uu}^2(t,x,y) + r\Big[M_{ou}^2(t,x+t-1,y) \\ &\quad + M_{uo}^2(t,x,y+t-1)\Big] \\ &- r\Big[M_{oou}^3(t,x+t-1,x+t-1,y) \\ &\quad + M_{uoo}^3(t,x,y+t-1,y+t-1)\Big], \\ &\text{for} \quad x, y \in (1-t, 1]. \end{aligned} \tag{5.63}$$

To solve these equations, one needs to solve first for the moments M_{ou}^2, M_{uo}^2, and M_{oou}^3 which satisfy equations of the following form:

$$\frac{\partial M^2_{ou}(t,x,y)}{\partial t} = -\alpha M^2_{ou}(t,x,y) + \beta M^2_{oo}(t,x,y), \tag{5.64}$$

$$\frac{\partial M^3_{oou}(t,x,y,z)}{\partial t} = -\alpha M^3_{oou}(t,x,y,z) + r M^3_{ooo}(x,y,z+t-1) - r M^4_{oooo}(x,y,z+t-1,z+t-1). \tag{5.65}$$

Hence, the moments can be obtained by successively solving ordinary or hyperbolic partial differential equations. Suppose for illustration that first order moments of the initial measure are real positive constants:

$$M^1_o = m_1 \tag{5.66}$$

$$M^2_{oo} = m_2 \tag{5.67}$$

$$M^3_{ooo} = m_3 \tag{5.68}$$

$$M^4_{oooo} = m_4. \tag{5.69}$$

First Moment

For $M^1_u(t,x)$, the evolution equation (5.59) reduces to

$$\frac{\partial M^1_u(t,x)}{\partial t} = -\alpha M^1_u(t,x) + r(m_1 - m_2),$$

whose solution is

$$M^1_u(t,x) = \gamma_1 + \left[M^1_u(0,x) - \gamma_1\right] e^{-\alpha t} \quad \text{where } \gamma_1 \equiv \frac{r(m_1 - m_2)}{\alpha}.$$

At $t = 0$, from (5.31) and (5.32) we know that $v(1) = u_v(1)$. In addition, $M^1_o(t,x) \equiv M^1_v(t,x)$. Therefore, from (5.45) and (5.46),

$$M^1_o(t=0, x=1) = \int_C v(1) d\mu_0 = \int_C u_v(1) d\mu_0(v) = M^1_u(t=0, x=1),$$

and from the initial condition (5.46) we conclude $M^1_u(t=0,x) = m_1$. Hence

$$M^1_u(t,x) = \gamma_1 + [m_1 - \gamma_1] e^{-\alpha t}.$$

Second Moments

To obtain expressions for M^2_{uv}, M^2_{vu}, M^2_{uu} we have to solve their respective equations of motion (remember that M^2_{vv} is given). We first tackle (5.62) (this choice is arbitrary; (5.61) can be dealt with in the same manner):

$$\frac{\partial M^2_{uv}(t,x,y)}{\partial t} = \frac{\partial M^2_{uv}(t,x,y)}{\partial y} - \alpha M^2_{uv}(t,x,y) + r(m_2 - m_3) \tag{5.70}$$

with initial condition $M^2_{uv}(0,x,y) = M^2_{vv}(0,x,y) \equiv m_2$ for all x, y in the domains defined in (5.42). This initial condition is, as for the first moment, obtained from Eqs. (5.45) and (5.46). Equation (5.70) is solved using the method of characteristics, and the solution is

$$M^2_{uv}(t,x,y) = \gamma_2 + [m_2 - \gamma_2]\, e^{-\alpha t} \quad \text{where } \gamma_2 \equiv \frac{r(m_2 - m_3)}{\alpha}.$$

The moment $M^2_{vu}(t,x,y)$ can be obtained in a similar way and the result is

$$M^2_{vu}(t,x,y) = M^2_{uv}(t,x,y).$$

This equality is due to the fact that the moments of the initial measure are constant. Finally, it is necessary to solve (5.64) and (5.65) before obtaining M^2_{uu}. Using (5.67)–(5.69),

$$\begin{aligned}
M^2_{ou} &= \gamma_2 + [m_2 + \gamma_2]\, e^{-\alpha t} \\
M^2_{uo} &= \gamma_2 + [m_2 + \gamma_2]\, e^{-\alpha t} \\
M^2_{oou} &= \gamma_3 + [m_3 + \gamma_3]\, e^{-\alpha t} \quad \text{where } \gamma_3 \equiv \frac{r(m_3 - m_4)}{\alpha} \\
M^2_{uoo}(t) &= \gamma_3 + [m_3 + \gamma_3]\, e^{-\alpha t},
\end{aligned}$$

so that the evolution equation for M^2_{uu} becomes

$$\frac{\partial M^2_{uu}(t,x,y)}{\partial t} = -2\alpha M^2_{uu}(t,x,y) + 2r\gamma_2 + 2r\gamma_3 + 2re^{\alpha t}\, [m_2 - m_3 + \gamma_2 - \gamma_3].$$

This is a linear first order ordinary delay equation which can be solved to give

$$M^2_{uu}(t) = \frac{-2re^{-\alpha t}}{3\alpha}\, [m_2 - m_3 + \gamma_2 - \gamma_3] - \frac{r}{\alpha}\, [\gamma_2 - \gamma_3] + \mathcal{K} e^{2\alpha t},$$

where

$$\mathcal{K} \equiv \frac{2r}{3\alpha}\left(m_2 - m_3 + \frac{1}{2}(\gamma_3 - \gamma_2)\right) + m_2.$$

This analysis can be carried out in a similar way when the moments are not constants, but such that the equations derived above remain solvable analytically.

5.2.5 Invariant Measures

It is of physical interest to investigate the constraint to be satisfied by a measure μ, invariant under the action of a differential delay equation. For the nonlinear differential delay equation (5.57), the characteristic functional $\mathcal{Y}$ of such a measure is defined as

$$\mathcal{Y}[J_1] = \int_C \exp\left[i \int_0^1 J_1(x) u_v(x+t)\, dx \right] d\mu,$$

and so we have

$$\mathcal{Y}[J_1] = \mathcal{Z}_t[J_1, 0] \qquad \text{for all } \; t,$$

where $\mathcal{Z}_t[J_1, J_2]$ is given by (5.35). The Hopf equation (5.44) becomes

$$\begin{aligned} 0 = &\int_0^{1-t} J_1(x) \frac{\partial}{\partial x} \left(\frac{\delta \mathcal{Y}}{\delta J_1(x)} \right) dx - \alpha \int_{1-t}^1 J_1(x) \frac{\delta \mathcal{Y}}{\delta J_1(x)}\, dx \\ &+ r \int_{1-t}^1 J_1(x) \frac{\delta \mathcal{Y}}{\delta J_1(x+t-1)}\, dx \\ &- r i^{-1} \int_{1-t}^1 J_1(x) \frac{\delta^2 \mathcal{Y}}{\delta J_1^2(x+t-1)}\, dx, \quad \forall t. \end{aligned}$$

By choosing $t = 0$, the first integral in the Hopf equation must vanish so that we have

$$\frac{\partial}{\partial x} \left(\frac{\delta \mathcal{Y}}{\delta J_1(x)} \right) = 0 \quad \text{a.e.}$$

From this relation a *necessary* condition for the invariant measure follows: using (5.48), the moments must satisfy

$$\sum_{k=1}^{n} \frac{\partial}{\partial x_k} M^n_{u^q v^{(n-q)}}(x_1, \cdots, x_k) = 0.$$

5.2.6 The Hopf Equation and the Kramers–Moyal Expansion

Our treatment of delayed dynamics is developed in the spirit of statistical mechanics and ergodic theory. One of the powerful tools of modern statistical mechanics is the use of equations describing the evolution of densities of initial conditions under the action of a finite-dimensional dynamical system. When that system is a set of ordinary differential equations the evolution equation is known as the *generalized Liouville equation*. When the system is a set of stochastic differential equations perturbed by realizations of a Wiener process, the evolution of densities is given by the *Fokker–Planck equation*. In general, for finite-dimensional systems the evolution of densities is governed by the *Kramers–Moyal (K–M) equation*. It is of some interest to understand how the K–M formalism carries over to systems with an infinite number of degrees of freedom such as differential delay equations.

The Hopf equation is probabilistic in the sense that it describes a set of differential delay equations in the same way that the Schrödinger equation describes a microscopic physical system.[1] Given that this is precisely the role of the K–M expansion for finite-dimensional dynamical systems, it is important to clarify the relation between the functional version of the K–M equation and the Hopf-like equation (5.41) derived here. Our derivation of the functional version of the K–M expansion is inspired by the derivation of the n-dimensional case given by Risken (1984).

To make the connection between $\mathcal{Z}_t$ and the K–M expansion more explicit, consider the expansion (5.47) of the characteristic functional $\mathcal{Z}_t$. Using Eq. (5.48), (5.47) becomes

$$\mathcal{Z}_t[J_1, J_2] = \sum_{p=0}^{\infty}\sum_{q=0}^{p}\int_0^1 \cdots \int_0^1 \frac{i^p}{p!} M^p_{u^q v^{(p-q)}}(t, \mathbf{x}) \times \left(\prod_{j=1}^{q} J_1(x_j)\, dx_j\right)\left(\prod_{j=q+1}^{p} J_2(x_j)\, dx_j\right). \tag{5.71}$$

Let $\mathcal{P}[G, H|G', H']$ be the transition probability functional that given the pair $(G'(x+t), H'(x+t))$ in $C \times C$, with $x \in [0, 1-t] \times (1-t, 1]$ we obtain the pair $(G(x+t+t_*), H(x+t+t_*))$ for $x \in [0, 1-(t+t_*)] \times (1-(t+t_*), 1]$ (i.e., G (G') is an initial function which generates a solution H (H')).

$\mathcal{W}[v, S_{(t+t_*)}v]$ is related to the transition probability functional $\mathcal{P}$ by

$$\mathcal{W}[v, S_{(t+t_*)}v] = \int_C \mathcal{P}[v, S_{(t+t_*)}v|v', S_t v']\mathcal{W}[v', S_t v']\, d\mu_0(v'). \tag{5.72}$$

In addition, $\mathcal{W}$ is the inverse Fourier transform of the characteristic functional $\mathcal{Z}_t[J_1, J_2]$ introduced in (5.35)

$$\mathcal{W}[v, S_t v] = \int_C \Upsilon^{-1}\, \mathcal{Z}_t[J_1, J_2]\, d\mathcal{V}[J_1, J_2]. \tag{5.73}$$

---

[1]A Schrödinger equation can *always* be transformed into a Fokker–Planck equation, which is just a truncation of the K–M expansion, but the physical interpretation of the transformation remains unclear (Risken 1984).

Also,

$$\mathcal{Z}_{t+t_*} = \int_C \tilde{\Upsilon}^{-1} \mathcal{P}[v, S_{(t+t_*)}v|v', S_t v'] d\mu_0(v'),$$

and therefore,

$$\mathcal{P}[v, S_{(t+t_*)}v|v', S_t v'] = \int_C \tilde{\Upsilon} \mathcal{Z}_{t+t_*}[J_1, J_2] d\mathcal{V}[J_1, J_2], \tag{5.74}$$

where

$$\tilde{\Upsilon} = \exp\Big\{ -i\Big[\int_0^1 J_2(x)(v(x) - v'(x))dx + \int_0^1 J_1(x)\,(S_t v'(x) - S_{(t+t_*)}v(x))dx \Big]\Big\},$$

and the measure of integration $\mathcal{V}[J_1, J_2]$ is a measure like the one used in the definition (5.33) of the characteristic functional. More precisely *the measure $\mathcal{W}$ describes the distribution of functions in C generating pairs $(v, S_t v)$ under the action of the transformation* (5.34), *and the measure $\mathcal{V}[J_1, J_2]$ describes the distribution of functions generating pairs (J_1, J_2) in the same space*. Inserting (5.71) into Eq. (5.74) we obtain

$$\mathcal{P}[v, S_{(t+t_*)}v|v', S_t v'] = \int_C \tilde{\Upsilon} \Bigg[\sum_{p=0}^{\infty} \sum_{q=0}^{p} \int_0^1 \cdots \int_0^1 \frac{1}{p!} M^p_{u^q v^{(p-q)}}(t + t_*, \mathbf{x}) \times \left(\prod_{j=1}^{q} i J_1(x_j)\, dx_j \right) \times \left(\prod_{j=q+1}^{p} i J_2(x_j)\, dx_j \right) \Bigg] d\mathcal{V}[J_1, J_2]. \tag{5.75}$$

The Dirac δ functional is a straightforward generalization of the more usual N-dimensional version. It satisfies

$$\int_C \delta[H - G] d\omega = \begin{cases} 1 \text{ if } H = G \text{ almost everywhere} \\ 0 \text{ otherwise,} \end{cases}$$

where H and G are elements of C, ω is a measure defined on C, and the result of the integration is a number (not a function). We will use this definition to simplify

The above analysis is not restricted to delay differential equations of the form (5.28) and (5.29). The only real constraint imposed on the dynamical system under consideration is that its phase-space be a normed function space. Therefore, this analysis is also valid for the statistical investigation of partial differential equations. In fact this approach was pioneered by Hopf (1952) who derived an evolution equation for the characteristic functional describing the solutions of the Navier–Stokes equations statistically.

From (5.73), it is clear that taking the Fourier transform of (5.83) yields the evolution equation for the characteristic functional $\mathcal{Z}_t[J_1, J_2]$. However, in Sect. 5.2.1, we derived the Hopf evolution equation (5.41) for $\mathcal{Z}_t[J_1, J_2]$. Thus, we conclude that the Hopf equation (5.41) is the Fourier transform of the infinite-dimensional extension of the Kramers–Moyal expansion (5.83). From (5.81), the K–M coefficients are given by solving the partial differential equations presented in Sect. 5.2.4.

5.3 Discussion and Conclusions

The introduction of the joint characteristic functional (5.33) provides a tool for the investigation of differential delay equations from a probabilistic point of view. This approach is meaningful from a physical perspective when dealing with large collections of entities whose dynamics are governed by differential delay equations. Moreover, a probabilistic description is clearly needed when the models are formulated as stochastic differential delay equations. In this case, the characteristic functional (5.33) is no longer valid, but it can be modified in a way similar to that presented in Section 7 of Lewis and Kraichnan (1962) (in the context of stochastic partial differential equations), and a three-interval characteristic functional should then be considered.

Note that the expansion (5.47) is similar to functional expansions in quantum field theory and statistical mechanics which are treated in a perturbative manner and analyzed with Feynman diagrams. Although the moment partial differential equations of Sect. 5.2.4 can indeed be deduced from a graphical analysis of expansion (5.47) (see Losson 1991), it remains to be seen whether the introduction of proper Feynman diagrams might provide, through graphical manipulations, significant insight into delayed dynamics.

In this chapter we have derived a Hopf-like functional equation for the evolution of the characteristic functional of a measure defined on the space of initial functions for a class of nonlinear differential delay equations. Using perturbation theory, the Hopf equation is reduced to an infinite chain of partial differential equations for the moments of the evolving distributions of functions. The first two moments are obtained explicitly for a delay differential equation with a quadratic nonlinearity, when the moments of the initial differential delay equations measure are constant. Finally, we show that the Hopf equation is the Fourier transform of the infinite-dimensional version of the Kramers–Moyal expansion.

Chapter 6
The Method of Steps

Consider the augmented differential delay equation initial value problem (with $\tau \equiv 1$)

$$x'(t) = \begin{cases} \mathcal{G}\big(x(t)\big) & t \in [0, 1) \\ \mathcal{F}\big(x(t), x(t-1)\big) & t \geq 1 \end{cases} \tag{6.1}$$
$$x(0) = x_0,$$

with $x(t) \in \mathbb{R}$, and suppose that an ensemble of initial values x_0 is specified with density f_0. We would like to derive an evolution equation for the density $f(x, t)$ of the corresponding ensemble of solutions $x(t)$.

There is an important preliminary observation to be made. Ideally, as noted in Chap. 1, we would like to derive an evolution equation of the form

$$\frac{df}{dt} = \{\text{some operator}\}(f). \tag{6.2}$$

However, f cannot satisfy such an equation. This is because the family of solution maps $\{S_t\}$ for Eq. (6.1) does not form a semigroup.That is, the density f cannot be sufficient to determine its own evolution, as in (6.2), because the values $x(t)$ in the ensemble it describes are insufficient to determine *their* own evolution. This difficulty arises because f does not contain information about the past states of the ensemble, which is necessary to determine the evolution of the ensemble under (6.1). Thus, any solution to the problem must take a form other than (6.2).

J. Losson et al., *Density Evolution Under Delayed Dynamics*, Fields Institute Monographs 38, https://doi.org/10.1007/978-1-0716-1072-5_6

6.1 Ordinary Differential Equation System

The method of steps is sometimes used to write a differential delay equation as a system of ordinary differential equations. This is a promising connection, as we already know how densities evolve for ordinary delay equations (cf. Sect. 2.3 (page 11)).

6.1.1 *Method of Steps*

Let $x(t)$ be a solution of (6.1) and define for $n = 0, 1, 2, \ldots$ the functions

$$y_n(s) = x(n+s), \quad s \in [0, 1]. \tag{6.3}$$

Figure 6.1 illustrates this relationship between the y_n and x.

Since $x(t)$ satisfies (6.1), it follows that for $n \geq 1$,

$$y_n'(s) = \mathcal{F}\big(y_n(s), y_{n-1}(s)\big), \quad n = 1, 2, \ldots, \tag{6.4}$$

and y_0 satisfies

$$y_0'(s) = \mathcal{G}(y_0(s)). \tag{6.5}$$

Thus, the augmented differential delay equation becomes a system of evolution equations for the y_n, together with the set of compatibility or boundary conditions

$$\begin{aligned} y_0(0) &= x_0 \\ y_n(0) &= y_{n-1}(1), \quad n = 1, 2, \ldots \end{aligned} \tag{6.6}$$

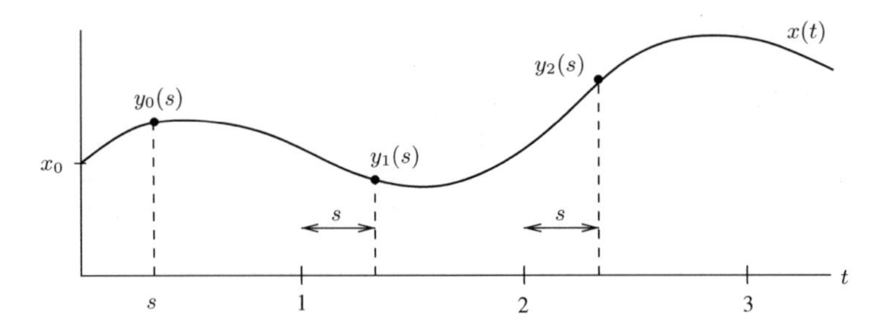

Fig. 6.1 Relationship between the differential delay equation solution $x(t)$ and the variables $y_n(s)$ defined in the method of steps, Eq. (6.3)

The ordinary delay equation system (6.4) and (6.5), together with these compatibility conditions, can be solved sequentially to yield the solution $x(t)$ of the differential delay equation up to any finite time. This is essentially the method of steps for solving the differential delay equation (cf. Sect. 4.2.2 (page 31)).

If the system of ordinary delay equations (6.4) and (6.5) could be taken together as a vector field $\tilde{\mathcal{F}}$ in $\mathbb{R}^{N+1}$, then an ensemble of solutions of the differential delay equation could be represented via (6.3) as an ensemble of vectors $y = (y_0, \ldots, y_N)$, each carried along the flow induced by $\tilde{\mathcal{F}}$. The density $\tilde{f}(y, t)$ of such an ensemble would evolve according to a continuity equation (cf. Sect. 2.3)

$$\frac{\partial \tilde{f}}{\partial t} = -\nabla \cdot (\tilde{f}\tilde{\mathcal{F}}).$$

However, it is unclear how to ensure the compatibility conditions (6.6) are satisfied by every vector in the ensemble, or how to determine the initial $(N+1)$-dimensional density $\tilde{f}(y, 0)$ of this ensemble in terms of a given density of initial values x_0 in (6.1). In short there is no obvious way to treat equations (6.4) and (6.5) simultaneously rather than sequentially. The following modified setup is one way to avoid these difficulties.

6.1.2 Modified Method of Steps

Any solution of the differential delay equation problem (6.1) can be extended unambiguously to all $t < 0$ by setting

$$x(t) = x_0, \quad t < 0,$$

so that $x'(t) = 0$ for all $t < 0$. For $n = 0, 1, 2, \ldots$ let functions y_n be defined by

$$y_n(t) = x(t - n), \quad t \geq 0. \tag{6.7}$$

Figure 6.2 illustrates the relationship between the y_n and x. On substitution into Eq. (6.1) we find that for $t \in [m, m+1]$, $m = 0, 1, 2, \ldots$, the $y_n(t)$ satisfy

$$y_n' = \begin{cases} \mathcal{F}(y_n, y_{n+1}) & \text{if } n < m \\ \mathcal{G}(y_m) & \text{if } n = m \\ 0 & \text{if } n > m. \end{cases}$$

Thus, for fixed N the vector $y(t) = \big(y_0(t), \ldots, y_N(t)\big)$ satisfies an ordinary differential equation

$$y' = \tilde{\mathcal{F}}(y, t), \tag{6.8}$$

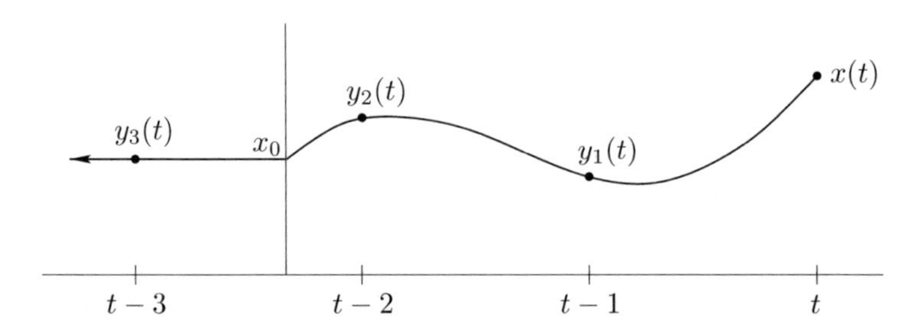

Fig. 6.2 Relationship between the differential delay equation solution $x(t)$ and the variables $y_n(t)$ defined in the modified method of steps, Eq. (6.7)

where the vector field $\tilde{\mathcal{F}} : \mathbb{R}^{N+1} \times \mathbb{R} \to \mathbb{R}^{N+1}$ is given by

$$\tilde{\mathcal{F}}(y_0, \ldots, y_N, t) = (y_0', \ldots, y_N'),$$
$$\text{with} \quad y_n' = \begin{cases} 0 & \text{if } t < n \\ \mathcal{G}(y_n) & \text{if } t \in [n, n+1) \\ \mathcal{F}(y_n, y_{n+1}) & \text{if } t \geq n+1. \end{cases} \tag{6.9}$$

Remark Some remarks about the ordinary delay equation system (6.8) and (6.9) are in order:

- For fixed $N > 0$, the right-hand side $\tilde{\mathcal{F}}(y, t)$, and hence the solution $y(t)$, is defined only for $0 \leq t \leq N + 1$.
- For any given initial vector $y(0) = (y_0(0), \ldots, y_N(0))$, Eq. (6.8) has a unique solution $y(t)$ defined for $0 \leq t \leq N + 1$, provided we have existence and uniqueness for the original differential delay equation problem (6.1).
- $\tilde{\mathcal{F}}(y, t)$ is piecewise constant in time. That is, for each $m = 0, 1, 2, \ldots$, the vector field $\tilde{\mathcal{F}}(y, t) = \tilde{\mathcal{F}}(y)$ is independent of $t \in [m, m + 1)$, and induces a flow in $\mathbb{R}^{N+1}$ that carries the solution $y(t)$ forward from $t = m$ to $t = m + 1$. Thus, we can speak of (6.8) as defining a sequence of flows in $\mathbb{R}^{N+1}$.

The ordinary delay equation system (6.8) gives a representation of the method of steps as an evolution equation in $\mathbb{R}^{N+1}$. Indeed, the solution $x(t)$ of the differential delay equation is given, up to time $t = N + 1$, by $x(t) = y_0(t)$ where $y(t)$ is the solution of (6.8) corresponding to the initial condition $y(0) = (x_0, \ldots, x_0)$. Thus, the differential delay equation problem (6.1) is equivalent to the initial value problem

$$\begin{aligned} &y' = \tilde{\mathcal{F}}(y, t), \quad y(t) \in \mathbb{R}^{N+1}, \ 0 \leq t \leq N + 1, \\ &y(0) = (x_0, \ldots, x_0), \end{aligned} \tag{6.10}$$

with the identification $x(t) = y_0(t)$.

6.2 Continuity Equation

Having established the equivalence of the differential delay equation system (6.1) with the ordinary delay equation system (6.10), we can proceed to the probabilistic treatment of differential delay equations using techniques developed for ordinary delay equations.

Suppose an ensemble of initial vectors $y \in \mathbb{R}^{N+1}$ is given, with $(N+1)$-dimensional density $\tilde{f}(y,t)$. Then under the sequence of flows induced by the vector field $\tilde{\mathcal{F}}(y,t)$, this density evolves according to a continuity (Liouville) equation (cf. Sect. 2.3),

$$\frac{\partial \tilde{f}(y,t)}{\partial t} = -\nabla \cdot \big(\tilde{f}(y,t)\tilde{\mathcal{F}}(y,t)\big), \tag{6.11}$$

where $\nabla = (\partial/\partial y_0, \ldots, \partial/\partial y_N)$. The initial density $\tilde{f}(y,0)$ derives from the density f_0 of initial values x_0 for the differential delay equation (6.1). That is, an ensemble of initial values x_0 with density f_0 corresponds to an ensemble of initial vectors $y = (x_0, \ldots, x_0)$ in $\mathbb{R}^{N+1}$, with "density"

$$\tilde{f}(y_0, \ldots, y_N; 0) = f_0(y_0)\delta(y_0 - y_1)\delta(y_0 - y_2)\cdots\delta(y_0 - y_N), \tag{6.12}$$

where δ is the Dirac delta function. This corresponds to a line mass concentrated on the line $y_0 = y_1 = \cdots = y_N$ with linear density $f_0(y_0)$. Thus, the problem of density evolution for delay differential equations becomes a problem of determining how this line mass is redistributed by the flow induced by (6.8).

With singular initial data such as (6.12), strong solutions of the continuity equation (6.11) do not exist. However, (6.11) can be interpreted in a weak sense that makes it possible to define "solutions" that satisfy initial conditions like (6.12). Such a weak solution can be obtained using the method of characteristics.

6.3 Method of Characteristics

Consider the initial value problem

$$\begin{aligned} \frac{\partial \tilde{f}(y,t)}{\partial t} + \nabla \cdot \big(\tilde{\mathcal{F}}(y,t)\tilde{f}(y,t)\big) = 0, \quad t \geq 0, \\ \tilde{f}(y,0) = g(y) \end{aligned} \tag{6.13}$$

for the unknown function $\tilde{f}(y,t)$, with $y = (y_0, \ldots, y_N)$. Assume provisionally that $\tilde{\mathcal{F}}$ and $\tilde{f}$ are differentiable in y, so the divergence operator can be expanded (by the product rule) to yield

$$\frac{\partial \tilde{f}}{\partial t} + \tilde{f}\nabla \cdot \tilde{\mathcal{F}} + \tilde{\mathcal{F}} \cdot \nabla \tilde{f} = 0. \tag{6.14}$$

Let a curve $\Gamma \subset \mathbb{R}^{N+2}$ (that is, in $(y_0, \ldots, y_N, t)$-space) be parametrized by smooth functions

$$y_0 = y_0(t), \ \ldots \ , y_N = y_N(t)$$

defined for all $t \geq 0$, and parametrize the value of $\tilde{f}$ on Γ by

$$\tilde{f}(t) = \tilde{f}\big(y_0(t), \ldots, y_N(t), t\big). \tag{6.15}$$

(This slight abuse of notation helps clarify the following development.) Differentiating (6.15) yields

$$\frac{d\tilde{f}}{dt} = \nabla \tilde{f} \cdot \frac{dy}{dt} + \frac{\partial \tilde{f}}{\partial t}.$$

Thus, if the functions $y(t)$, $\tilde{f}(t)$ satisfy

$$\frac{dy}{dt} = \tilde{\mathcal{F}}\big(y(t), t\big) \tag{6.16}$$

$$\frac{d\tilde{f}}{dt} = -\tilde{f}(t)\nabla \cdot \tilde{\mathcal{F}}\big(y(t), t\big) \tag{6.17}$$

for all $t \geq 0$, then $\tilde{f}$ as given by Eq. (6.15) satisfies the partial differential equation (6.14) at every point on Γ. In fact any solution of the ordinary delay equation system (6.16) and (6.17) furnishes a solution of the partial differential equation (6.14) on a particular curve Γ. In particular, if y, $\tilde{f}$ are solutions of this system corresponding to initial values

$$y(0) = (y_0, \ldots, y_N)$$

$$\tilde{f}(0) = g\big(y_0, \ldots, y_N\big),$$

then Γ intersects the hyperplane $t = 0$ at the point $y = (y_0, \ldots, y_N)$, where $\tilde{f}(0)$ agrees with the initial data given by g, so that $\tilde{f}(t)$ gives the solution of (6.13) at every point of Γ.

The family of curves Γ that satisfy (6.16) are called the *characteristics* of the partial differential equation (6.14). As we have seen, the characteristics have the geometric interpretation that an initial datum specified at $(y_0, \ldots, y_N, 0)$ is propagated along the characteristic that passes through this point. Not surprisingly, the characteristic curves of (6.13) coincide with the integral curves of the vector field $\tilde{\mathcal{F}}$ (cf. Eq. (6.16)). That is, initial data are propagated along streamlines of the induced flow.

If the characteristics foliate $\mathbb{R}^{N+2}$, every point $(y_0, \ldots, y_N, t) \in \mathbb{R}^{N+2}$ has a characteristic curve passing through it. Then the solution of (6.13) can be found at any point, by using (6.16) and (6.17) to obtain the solution on the characteristic curve through that point. The usual procedure for obtaining the solution $\tilde{f}(y, t)$ at $P = (y_0, \ldots, y_N, t)$ is as follows.

1. Determine the characteristic curve Γ through P and follow it "backward" in time to find the point Q on Γ at $t = 0$.
2. Evaluate g at Q to determine the initial value $\tilde{f}(0)$ on Γ.
3. With this value of $\tilde{f}(0)$, integrate equation (6.17) forward along Γ to P, at which point the value of $\tilde{f}(t)$ is the solution of (6.13).

Figure 6.3 gives a schematic illustration of the method.

Although the derivation assumes differentiability of $\tilde{f}$ (hence also of g), the method itself does not rely on any special properties of these functions—it requires only integration of the vector field F and evaluation of ϕ. Hence the method can be applied even if g is discontinuous, or singular as in (6.12). However, in such cases the resulting function $\tilde{f}$ must be interpreted as a weak solution (Carrier and Pearson 1988; Zauderer 1983).

Supposing the solution $\tilde{f}(y, t)$ of the initial value problem (6.11) and (6.12) to have been found—e.g., by the method of characteristics—the corresponding density $f(x, t)$ of differential delay equation solutions $x(t)$ can be determined as follows. Since $x(t) = y_0(t)$ (cf. Eq. (6.7)), the density of x is identified with the density of y_0. This density is determined by integrating $\tilde{f}$ over all components except y_0, i.e.,

$$f(y_0, t) = \int \cdots \int \tilde{f}(y_0, y_1, \ldots, y_N, t)\, dy_1 \cdots dy_N.$$

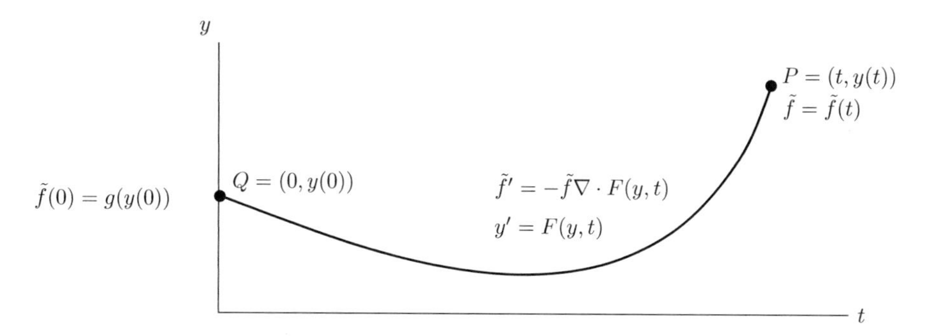

Fig. 6.3 Method of characteristics for the continuity equation (6.13): propagation of initial data along an integral curve of $y' = F(y, t)$

6.3.1 Alternative Formulation

There is another way to formulate the evolution of the density $\tilde{f}(y,t)$ that turns out to be equivalent to the method of characteristics and serves to illuminate the method above. It also provides an explicit formula (actually a codification of the algorithm on page 85) for the solution $\tilde{f}(y,t)$.

Recall that the vector $y(t)$ evolves according to a system of ordinary delay equations (6.8) and (6.9). Let $y(t)$ be the solution of this system with initial value $y(0)$, and define the corresponding family of solution maps $\hat{S}_t : \mathbb{R}^{N+1} \to \mathbb{R}^{N+1}$ by

$$\hat{S}_t : y(0) \mapsto y(t),$$

to be distinguished from the solution map S_t for the differential delay equation. As $y(t)$ evolves under the action of $\hat{S}_t$, the density $\tilde{f}(y,t)$ evolves according to the corresponding Frobenius–Perron operator $P^t : \tilde{f}(y,0) \mapsto \tilde{f}(y,t)$, defined by

$$\int_A P^t \tilde{f}(y,0)\, d^{N+1}y = \int_{\hat{S}_t^{-1}(A)} \tilde{f}(y,t)\, d^{N+1}y \quad \forall \text{ Borel } A \subset \mathbb{R}^{N+1}. \tag{6.18}$$

Recall that $\hat{S}_t$ can be represented as a composition of flows on the intervals $[m, m+1]$, $m = 0, 1, 2, \ldots$, so it is one-to-one on $\mathbb{R}^{N+1}$ and has an inverse (which can be found by reversing the sequence of flows). This makes possible the change of variables $z = \hat{S}_t(y)$ in (6.18), which by Theorem 3.2.1 of Lasota and Mackey (1994) becomes

$$\int_A P^t \tilde{f}(y,0)\, d^{N+1}y = \int_A \tilde{f}\big(\hat{S}_t^{-1}(z), t\big) J_t^{-1}(z)\, d^{N+1}z.$$

Since A is arbitrary, this implies the following explicit formula for P^t:

$$P^t \tilde{f}(y,0) = \tilde{f}(y,t) = \tilde{f}\big(\hat{S}_t^{-1}(y), 0\big) J_t^{-1}(y). \tag{6.19}$$

Here J_t^{-1} is the density of the measure $m \circ \hat{S}_t^{-1}$ with respect to Lebesgue measure m (Lasota and Mackey 1994, p. 46). If $\hat{S}_t$ and $\hat{S}_t^{-1}$ are differentiable transformations[1] of $\mathbb{R}^{N+1}$, then J_t^{-1} is just the determinant of the Jacobian matrix $D\hat{S}_t^{-1}$,

$$J_t^{-1}(y) = \det\big(D\hat{S}_t^{-1}(y)\big),$$
$$J_t(y) = \det\big(D\hat{S}_t(y)\big).$$

[1]It suffices that the vector field $F(y,t)$ be smooth in y (Kuznetsov 1995, p. 19).

In this case the formula (6.19) is a multi-dimensional representation of the Frobenius–Perron operator when $\hat{S}_t$ is invertible.

Notice that $\hat{S}_t$ effects the translation of a point $(y(0), 0)$ along a characteristic curve Γ to $(y(t), t)$. Similarly, $\hat{S}_t^{-1}$ effects a translation backward along Γ to $t = 0$. This draws the connection between (6.19) and the method of characteristics (page 85): the point Q (where the initial density is evaluated) is identified with the point $(\hat{S}_t^{-1}(y), 0)$ in (6.19).

The factor $J_t^{-1}(y)$ also has a geometric interpretation: it is the factor by which the volume of an infinitesimal volume element at $y(t)$ increases under transportation by $\hat{S}_t^{-1}$. This factor can equivalently be understood as resulting from step 3 of the method of characteristics algorithm, since $\nabla \cdot \tilde{\mathcal{F}}(y, t)$ is the instantaneous growth rate of an infinitesimal volume at y as it is transported by the flow $\hat{S}_t$ induced by $\tilde{F}$. Conservation of mass requires that the density supported on an infinitesimal volume element decreases in proportion to the volume growth, i.e., by the factor $J_t^{-1}(y)$. This provides a geometrical explanation of the term $J_t^{-1}(y)$ in (6.19).

A number of examples illustrating this approach to the evolution of densities under the action of delayed dynamics can be found in Taylor (2004). However, as the examples suggest, for all but the simplest delay equations an analytical treatment of the density evolution problem is difficult, and perhaps impossible. Nevertheless, the approach developed above does provide some geometrical insights even when an analytical approach fails.

6.4 Geometric Interpretation

Recall that up to any finite time $t \leq N + 1$ the differential delay equation problem (6.8) and (6.9) can be represented by an ordinary differential equation

$$y' = \tilde{\mathcal{F}}(y, t), \quad y(t) \in \mathbb{R}^{N+1}, \quad t \geq 0, \tag{6.20}$$

with $\tilde{\mathcal{F}}$ defined by (6.9). An ensemble of initial values with density f_0 corresponds to an ensemble of initial vectors y with $(N + 1)$-dimensional "density"

$$\tilde{f}(y_0, \dots, y_N; 0) = f_0(y_0)\delta(y_0 - y_1)\delta(y_1 - y_2) \cdots \delta(y_{N-1} - y_N),$$

representing a line mass concentrated on the line $y_0 = y_1 = \cdots = y_N$ in $\mathbb{R}^{N+1}$. Under evolution by (6.20), i.e., under transformation by the solution map $\hat{S}_t$, this line mass is redistributed. This transportation of a line mass under $\hat{S}_t$ is illustrated in Fig. 6.4. After evolution by time t, $\tilde{f}(y, t)$ is supported on a one-dimensional curve that is the image of this line under $\hat{S}_t$. We will call this curve the "density support curve." It is a continuous, non-self-intersecting curve in $\mathbb{R}^{N+1}$, owing to continuity and invertibility of $\hat{S}_t$.

Figure 6.5 illustrates the results of applying this idea to the Mackey–Glass equation (3.1), for an ensemble of constant initial functions with values distributed

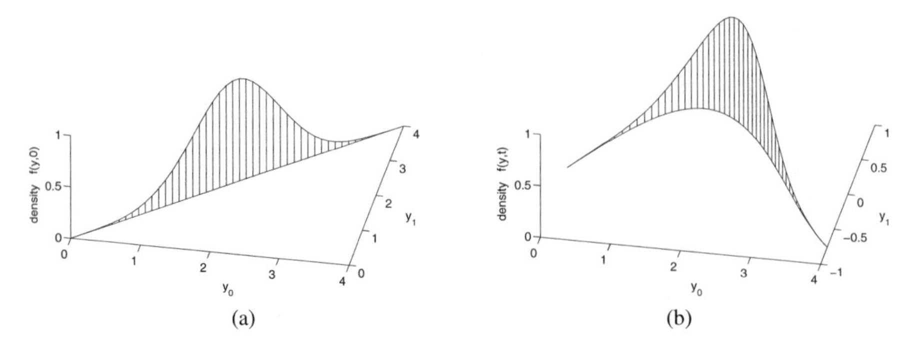

Fig. 6.4 Transport of a line mass under a transformation of the (y_0, y_1)-plane. (**a**) Initial mass distributed on the line $y_0 = y_1$. (**b**) Mass distribution after transformation

on the interval [0.3, 1.3], as in Figs. 3.2 and 3.3. Thus, the initial density support curve is the part of the line $y_0 = y_1 = \cdots$ with $0.3 \le y_0 \le 1.3$.

The first row of Fig. 6.5 shows the sequence of density support curves obtained at times $t = 1, 2, 3, 4$, projected onto the (y_0, y_1)-plane (as a result of this projection, some of the curves intersect themselves). The second row shows the corresponding densities $f(x, t)$ resulting from projecting the mass supported on the corresponding density support curve onto the y_0-axis. With this interpretation, the density support curves provide an obvious geometrical interpretation of the structures observed in the corresponding densities $f(x, t)$. Discrete jumps in the density occur at the endpoints of the transformed density support curve, and the maxima (singularities) correspond to turning points of the transformed density support curve.

For large t the transformed density support curve becomes very complicated. Figure 6.6 shows the density support curve at time $t = 20$, which follows in the sequence of Fig. 6.5. The complexity of this curve results from the repeated stretching and folding that occurs under the dynamics of the differential delay equation. Because of this complexity it is difficult to provide a clear geometric interpretation of the corresponding density, as was possible for small times as in Fig. 6.5. Also, just as with the solution map, determining the density support curve numerically becomes problematic (in fact impossible using finite precision) for large times.

6.5 Methods of Steps and Piecewise Deterministic Markov Processes

In this section we show how the system of Eqs. (6.4) and (6.5) with compatibility condition (6.6) can be embedded into a family of piecewise deterministic Markov processes as introduced by Davis (1984), see also Rudnicki and Tyran-Kamińska (2017). With such a process one can consider an increasing sequence of (random)

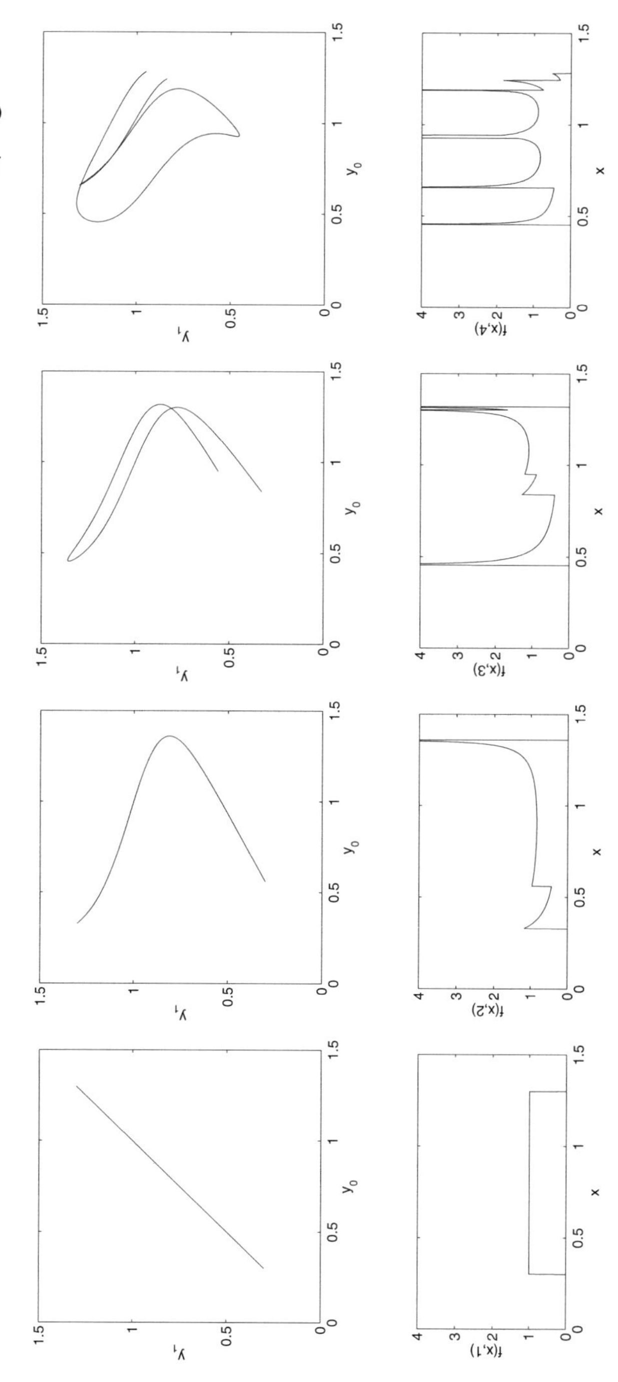

Fig. 6.5 Density support curves (projected onto the (y_0, y_1)-plane) for the Mackey–Glass equation (3.1) restricted to constant initial functions. The second row of graphs shows the corresponding computed densities

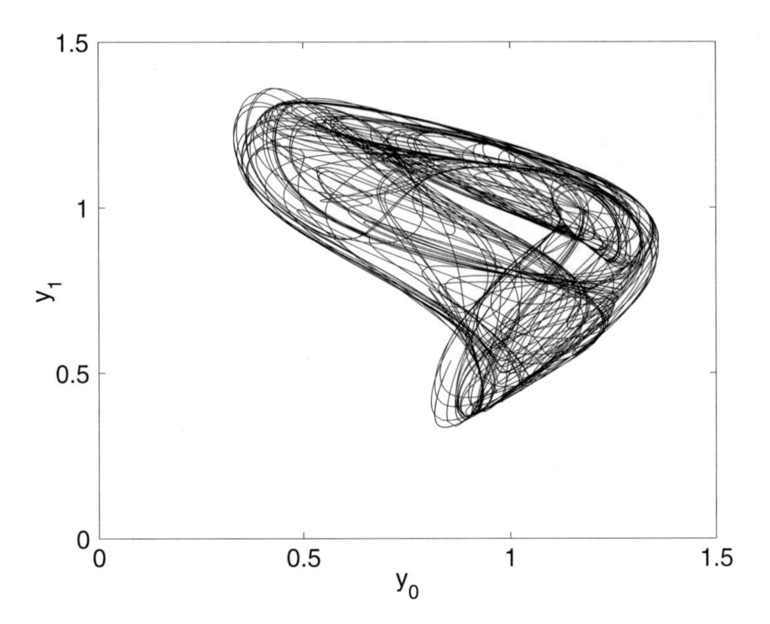

Fig. 6.6 Density support curve (projected onto the (y_0, y_1)-plane) for the Mackey–Glass equation (3.1) restricted to constant initial functions, evolved forward in time to $t = 20$

jump times $\tau_0, \tau_1, \tau_2, \ldots$ such that on each time interval (τ_n, τ_{n+1}) the process is deterministic and described by a flow ϕ_t, while at the jump times $\tau_0\tau_1, \tau_2, \ldots$ the values $X_0, X_1, X_2, \ldots$ of the process can be taken from some distribution or can be prescribed deterministically as well. The latter can happen in particular if the process hits the boundary of the state space in which case there is a forced jump from the boundary back to the state space. The process $\{X(t) : t \geq 0\}$ starting from X_0 at time $\tau_0 = 0$ is defined by

$$X(t) = \phi_{t-\tau_n}(X_n) \quad \text{if } \tau_n \leq t < \tau_{n+1}, n \geq 0. \tag{6.21}$$

Such processes and evolution equations for their densities were studied recently in Gwizdz and Tyran-Kamińska (2019).

Let us first introduce some notation connected with the system of Eqs. (6.4) and (6.5). For each $n \geq 0$ define

$$\mathrm{y}_n = (y_0, y_1, \ldots, y_n) \quad \text{and} \quad \mathcal{F}_n(\mathrm{y}_n) = (\mathcal{G}(y_0), \mathcal{F}(y_1, y_0), \ldots, \mathcal{F}(y_n, y_{n-1})). \tag{6.22}$$

Let for $\mathrm{x} \in \mathbb{R}^{n+1}$ the unique solution of the initial value problem

$$\mathrm{y}_n'(t) = \mathcal{F}_n(\mathrm{y}_n(t)), \quad \mathrm{y}_n(0) = \mathrm{x}, \tag{6.23}$$

be denoted by $S_t^n(\mathrm{x})$. The compatibility condition (6.6) can be rewritten as

$$\mathrm{y}_0(0) = x_0, \quad \mathrm{y}_n(0) = (x_0, \mathrm{y}_{n-1}(1)) = (x_0, S_1^{n-1}(\mathrm{y}_{n-1}(0))), \quad n \geq 1. \tag{6.24}$$

Note that we have

$$\mathrm{y}_1(0) = (x_0, S_1^0(x_0)), \quad \mathrm{y}_2(0) = (x_0, S_1^1(x_0, S_1^0(x_0)))$$

and so on. Thus, if we define

$$R_0(x_0) = x_0, \quad R_n(x_0) = (x_0, S_1^{n-1}(R_{n-1}(x_0))), \quad n \geq 1, \tag{6.25}$$

then we obtain $\mathrm{y}_n(0) = R_n(x_0)$, $n \geq 0$ implying that the general solution of the differential equation $\mathrm{y}_n'(s) = \mathcal{F}_n(\mathrm{y}_n(s))$ satisfying the compatibility condition (6.6) is given by

$$\mathrm{y}_n(s) = S_s^n(R_n(x_0)), \quad s \in [0, 1].$$

Since the kth coordinate of $S_t^n(y_0, \ldots, y_n)$ depends only on the variables $y_0, \ldots, y_k$, we may introduce

$$\psi_t^0(y_0) = S_t^0(y_0), \quad \psi_t^n(y_0, \ldots, y_n) = \pi_n S_t^n(y_0, \ldots, y_n), \quad n \geq 1,$$

where $\pi_n : \mathbb{R}^{n+1} \to \mathbb{R}$ is the projection $(y_0 \ldots, y_n) \mapsto y_n$. Thus we obtain

$$S_t^n(y_0, \ldots, y_n) = (\psi_t^0(y_0), \psi_t^1(y_0, y_1), \ldots, \psi_t^n(y_0, \ldots, y_n)). \tag{6.26}$$

Consequently, for $t \in [0, \infty)$ the solution of (6.1) is given by

$$x(t) = \psi_{t-n}^n(R_n(x_0)) \quad \text{if } t \in [n, n+1). \tag{6.27}$$

Let us now define the Markov process. For each $n \geq 0$ we consider the solution $\phi_t^n(s_0, \mathrm{x})$ of the system of equations

$$s'(t) = 1, \quad \mathrm{y}_n'(t) = \mathcal{F}_n(\mathrm{y}_n(t)), \quad s(0) = s_0, \mathrm{y}_n(0) = \mathrm{x}, \tag{6.28}$$

where $s_0 \in \mathbb{R}$ and $\mathrm{x} \in \mathbb{R}^{n+1}$. We have

$$\phi_t^n(s_0, \mathrm{x}) = (s_0 + t, S_t^n(\mathrm{x})), \tag{6.29}$$

where $S_t^n(x)$ denotes the solution of (6.23). Let

$$\tilde{E} = \bigcup_{n=0}^{\infty} \mathbb{R}^{n+2} \times \{n\}.$$

We define a flow ϕ_t on $\tilde{E}$ by

$$\phi_t(s, \mathrm{x}, n) = (\phi_t^n(s, \mathrm{x}), n), \quad (s, \mathrm{x}) \in \mathbb{R}^{n+2}, n \geq 0. \tag{6.30}$$

If we let $s = 0$ at time $t = 0$, then after time $t = 1$ we reach the point $\mathrm{y}_n(1) = S_1^n(\mathrm{x})$. Thus, we restrict the values of s to the interval $[0, 1]$ and for each n we consider the solution ϕ_t^n of Eq. (6.28) on the space $E_n = [0, 1) \times \mathbb{R}^{n+1}$. Once the boundary $\{1\} \times \mathbb{R}^{n+1}$ of E_n is reached by the solution flow ϕ_t^n, say at the point $(1, y_0, \ldots, y_n)$, then we increase n by 1 and we jump to the point $(0, S_{-1}^0(y_0), y_0, \ldots, y_n)$ that becomes the initial point for the flow ϕ_t^{n+1} in the space $E_{n+1} = [0, 1) \times \mathbb{R}^{n+2}$. After reaching the boundary $\{1\} \times \mathbb{R}^{n+2}$ of E_{n+1} by the flow ϕ_t^{n+1} we jump to the corresponding initial point for the flow ϕ_t^{n+2} and so on.

The state space of the Markov process $\{X(t) : t \geq 0\}$ as in (6.21) is taken to be

$$E = \bigcup_{n=0}^{\infty} E_n \times \{n\} = \bigcup_{n=0}^{\infty} [0, 1) \times \mathbb{R}^{n+1} \times \{n\}.$$

The flow ϕ_t defined in (6.29) and (6.30) can reach the boundary

$$\Gamma^+ = \bigcup_{n=0}^{\infty} \{1\} \times \mathbb{R}^{n+1} \times \{n\}$$

of the space E in a finite time $t_+(s, \mathrm{x}, n) = 1 - s$ for $s \in [0, 1)$, $n \in \mathbb{N}$, $\mathrm{x} \in \mathbb{R}^{n+1}$. The only possible jumps for the process are from the boundary Γ^+, so that from a point $(1, y_0, \ldots, y_n, n) \in \Gamma^+$ we go with probability one to the point $(0, S_{-1}^0(y_0), y_0, \ldots, y_n, n+1) \in \Gamma^-$, where

$$\Gamma^- = \bigcup_{n=0}^{\infty} \{n\} \times \{0\} \times \mathbb{R}^{n+1}.$$

Let $x_0 \in \mathbb{R}$ and $s_0 = 0$. The Markov process $X(t)$ starting from $X_0 = (0, x_0, 0)$ at time $\tau_0 = 0$ is given by

$$X(t) = \phi_t(0, x_0, 0) = (t, S_t^0(x_0), 0)$$

as long as $t < 1$. The moment τ_1 of the first jump is 1 and we have

$$X(1^-) = \lim_{t \to 1^-} X(t) = (1, S_1^0(x_0), 0).$$

Since $S_{-1}^0(S_1(x_0)) = x_0$, we jump from $X(1^-)$ to the point

$$X_1 = (0, x_0, S_1^0(x_0), 1).$$

For $t \geq 1$ we have

$$X(t) = \phi_{t-1}(X_1) = (t-1, S^1_{t-1}(x_0, S^0_1(x_0)), 1)$$

as long as $t - 1 < 1$. The moment τ_2 of the second jump is 2 and we have

$$X(2^-) = \lim_{t \to 2^-} X(t) = (1, S^1_1(x_0, S^0_1(x_0)), 1).$$

Thus we jump from $X(2^-)$ to the point

$$X_2 = (0, x_0, S^1_1(x_0, S^0_1(x_0)), 2),$$

and so on. Consequently, we obtain $\tau_n = n$ and

$$X(t) = (t-n, S^n_{t-n}(R_n(x_0)), n), \quad t \in [n, n+1),$$

where $R_n(x_0)$ is defined as in (6.25). We see that the solution $x(t)$ of (6.1), as given by (6.27), is recovered from $X(t)$ through $x(t) = \tilde{\pi}_n(X(t))$ for $t \in [n, n+1)$, where $\tilde{\pi}_n$ is the projection $(s, y_0, \ldots, y_n, n) \mapsto y_n$, $n \geq 0$.

We would like to find the probability densities of the Markov process. Let $\mathcal{E}$ be the σ-algebra which is the union of Borel σ-algebras of subsets of E_n. We can consider the σ-finite measure m on $\mathcal{E}$ given by

$$m = \sum_{n \in \mathbb{N}} m_{n+2} \times \delta_n,$$

where m_{n+2} is the Lebesgue measure on $\mathbb{R}^{n+2}$ and δ_n is the point measure on the set $\{n\}$. The densities, if they exist, belong to the space $L^1(E, \mathcal{E}, m)$ and satisfy the following system of equations

$$\frac{\partial u(s, y_0, 0, t)}{\partial t} = -\frac{\partial u(s, y_0, 0, t)}{\partial s} - \frac{\partial (\mathcal{G}(y_0) u(s, y_0, 0, t))}{\partial y_0} \tag{6.31}$$

and

$$\frac{\partial u(s, y_0, \ldots, y_n, n, t)}{\partial t} = -\frac{\partial u(s, y_0, \ldots, y_n, n, t)}{\partial s} - \frac{\partial (\mathcal{G}(y_0) u(s, y_0, \ldots, y_n, n, t))}{\partial y_0} - \sum_{i=1}^{n} \frac{\partial (\mathcal{F}(y_i, y_{i-1}) u(s, y_0, \ldots, y_n, n, t))}{\partial y_i}, \tag{6.32}$$

considered together with an initial density and with the boundary conditions $u(0, y_0, 0, t) = 0$ and

$$u(0, y_0, y_1, \dots, y_n, n, t) = \delta_{y_0}(S^0_{-1}(y_1))u(1, y_1, \dots, y_n, n-1, t), \quad n \geq 1. \tag{6.33}$$

However, the question of the existence of solutions to the initial-boundary value problem (6.31)–(6.33) remains open. The boundary conditions are different than those studied in Gwizdz and Tyran-Kamińska (2019) where the boundary sets Γ^+ and Γ^- are considered with measures

$$m^+ = \sum_{n\in\mathbb{N}} \delta_1 \times m_{n+1} \times \delta_n, \quad m^- = \sum_{n\in\mathbb{N}} \delta_0 \times m_{n+1} \times \delta_n$$

and the boundary conditions are written as

$$u_{|\Gamma^-} = P(u_{|\Gamma^+})$$

with $u_{|\Gamma^\pm}$ denoting restrictions of u to the boundaries $\Gamma^\pm$ and $P\colon L^1(\Gamma^+, m^+) \to L^1(\Gamma^-, m^-)$ being a positive linear operator.

We can also embed the delay differential equation (4.2) with $\tau = 1$ into a family of Markov processes by a slight modification of the above construction. Given $\phi \in C([-1, 0])$ we extend it to a continuous function on $[-1, \infty)$ by writing $\phi(t) = \phi(0)$ for $t \geq 0$. For each $n \geq 0$ we redefine $\mathcal{F}_n$ in (6.22) in the following way:

$$\mathcal{F}_n(\mathrm{y}_n, s) = (\mathcal{F}(y_0, \phi(s-1)), \mathcal{F}(y_1, y_0), \dots, \mathcal{F}(y_n, y_{n-1})),$$

and we let $\phi^n_t(s_0, \mathrm{x})$ be the solution of equation

$$s'(t) = 1, \quad \mathrm{y}'_n(t) = \mathcal{F}_n(\mathrm{y}_n(t), s(t)), \quad s(0) = s_0, \mathrm{y}_n(0) = \mathrm{x}.$$

As the state space of the process we now take the set

$$E = \bigcup_{n=0}^{\infty} C([-1, 0]) \times [0, 1) \times \mathbb{R}^{n+1} \times \{n\},$$

and we define the flow ϕ_t by

$$\phi_t(\phi, s, \mathrm{x}, n) = (\phi, \phi^n_t(s, \mathrm{x}), n), \quad (\phi, s, \mathrm{x}) \in C([-1, 0]) \times [0, 1) \times \mathbb{R}^{n+1}, n \geq 0.$$

Once the boundary of E is reached at a point $(\phi, 1, y_0, \dots, y_n, n)$ we jump to the point $(\phi, 0, \phi(0), y_0, \dots, y_n, n+1)$ with probability one. An analogous construction as above leads to the Markov process $\{X(t) : t \geq 0\}$ started from $(\phi, 0, \phi(0), 0)$. In the corresponding equations for densities (6.31) and (6.32) we need to have a dependence on ϕ so that $u = u_\phi$ and we change $\mathcal{G}(y_0)$ to $\mathcal{F}(y_0, \phi(s-1))$, while in the boundary condition (6.33) we change $S^0_{-1}(y_1)$ to $\phi(0)$. Here again, we are left with the problem of the existence of solutions to such equations.

6.6 Summary and Conclusions

In this chapter we have returned to a consideration of the method of steps as first introduced in Sect. 4.2.2, by introducing a modification of the technique, leading to an equivalent initial value problem (6.10). However, that seems to lead to a "weak solution." We continued this investigation using the method of characteristics that gives an alternative way of looking at the problem, and then provided a geometric interpretation using the Mackey–Glass equation as a test example. We also embedded the method of steps into a family of Markov processes.

Part IV
Possible Approximating Solutions

In this final section we examine two possible approaches in which alternative approximations are employed to turn first order differential delay equations into maps. Such similarities might enable us to analyze the dynamics of differential delay equations using the powerful mathematical tools that have been developed for maps.

The differential delay equation of interest is again (1.4)

$$\epsilon \frac{dx}{dt} = -x(t) + S(x(t-\tau)). \tag{IV.1}$$

The limit where $\epsilon \to 0$ gives

$$x(t) = S(x(t-\tau)), \tag{IV.2}$$

and Eq. (IV.1) is the singular perturbation problem corresponding to (IV.2). Equation (IV.2) can be viewed as a recipe for functional iteration, and if we knew how to define densities of initial functions, then we could discuss their evolution under the action of (IV.2) and might be some way toward understanding our original problem.

Denoting $x_n \equiv x(n\tau)$, (IV.2) can be written as a discrete time map:

$$x_{n+1} = S(x_n).$$

The dynamical behaviors of this map, such as period-doubling bifurcations and chaotic motion, are also found in the differential delay equation. In fact, the differential delay equation exhibits a much broader range of dynamical behaviors than the map obtained in the singular limit. In general, there is no continuous transition between the dynamical structures of the map (periodic orbits, their stability properties, bifurcation points) and those of the differential delay equation.

Chapter 7 (page 99) considers a high-dimensional map approximation to the delay equation in Sect. 7. Then, Chap. 8 (page 115) is devoted to developing approximate Liouville-like equations and an examination of invariant densities for differential delay equations.

Chapter 7
Turning a Differential Delay Equation into a High-Dimensional Map

This entire section was originally published in Losson and Mackey (1995). In this chapter we examine the potential use of a variety of approximations, or reductions, of a differential delay equation to a system of ordinary differential equations in the first instance, and reducing the delay differential equation to a high-dimensional map in the second.

As we have repeatedly pointed out, the phase space of differential delay equations is infinite-dimensional. The ensemble density, which gives the probability of occupation of phase space, is therefore a functional. The evolution equation for this functional, known as the Hopf equation (Capiński 1991; Hopf 1952; Lewis and Kraichnan 1962; Losson and Mackey 1992), cannot be integrated due to the lack of a theory of integration with respect to arbitrary functional measures. See in particular Sect. 5.2.

In this section we propose a reduction of the original differential delay equation to a finite-dimensional system which is arbitrarily accurate. The work presented here strongly indicates that in many circumstances of interest (from a modeling perspective) the Hopf equation can be approximated by the Frobenius–Perron equation in $\mathbb{R}^N$ (or its stochastic equivalent). The resulting description of delayed dynamics is akin to the description of ordinary differential equations given by the generalized Liouville equation, or of the Langevin equation by the Fokker–Planck equation. Once the reduction is completed, the analytical techniques available can then be used to explain the presence of continuous-time statistical cycling numerically observed in the differential delay equations.

J. Losson et al., *Density Evolution Under Delayed Dynamics*, Fields Institute Monographs 38, https://doi.org/10.1007/978-1-0716-1072-5_7

7.1 From Differential Delay to Ordinary Differential Equations

The link between hereditary dynamical systems (framed as functional or delay differential equations) and spatially extended models (hyperbolic partial differential equations to be precise) has been discussed extensively (cf. Blythe et al. 1984; Cooke and Krumme 1968; Vogel 1965). In a rather formal context, Fargue (1973, 1974) argues that it is possible to interpret hereditary systems as being nonlocal, or extended. This allows the introduction of a field which is intrinsic to the system, and the variable which satisfies the hereditary model is then a functional of this field. In other words, the memory in the system is interpreted as a nonlocality. At a more applied level, Sharkovskiĭ et al. (1986) have shown that systems of hyperbolic partial differential equations could, given appropriate boundary condition, be reduced via use of the method of characteristics to differential delay equations of the first order.

In this section we will assume that we have scaled the parameters in such a way that $\tau = 1$.

The differential delay equations considered in this section are of the form

$$\frac{dx(t)}{dt} = -\alpha x(t) + S(x(t-1)) \tag{7.1}$$

with an initial function $\phi(s)$ defined for $s \in [-1, 0)$. There is a continuous time semi-dynamical system associated with (7.1), given by

$$\frac{dx_\phi(t)}{dt} = \begin{cases} \dfrac{d\phi(t)}{dt} & \text{if } t \in [-1, 0) \\ -\alpha x_\phi(t) + S(x_\phi(t-1)) & \text{if } t \geq 0, \end{cases}$$

so that the differential delay equation (7.1) defines a continuous time operator S_t acting on bounded functions defined everywhere on $[-1, 0)$. For example, if ϕ denotes such an initial function,

$$S_t\phi = \{x_\phi(s) : s \in [t-1, t)\}, 0 \leq t \leq 1 \tag{7.2}$$

(if $t > 1$, the initial function is no longer ϕ).

The first step in the reduction of (7.1) is to use the Euler approximation to dx/dt and write

$$\lim_{\Delta \to 0} \frac{x_\phi(t) - x_\phi(t-\Delta)}{\Delta} = -\alpha x_\phi(t) + S(x_\phi(t-1)), \qquad \Delta > 0. \tag{7.3}$$

Removing the limit, (7.3) can be approximated by

$$x_\phi(t) = \frac{1}{(1+\alpha\Delta)} \left[x_\phi(t-\Delta) + \Delta S(x_\phi(t-1))\right], \tag{7.4}$$

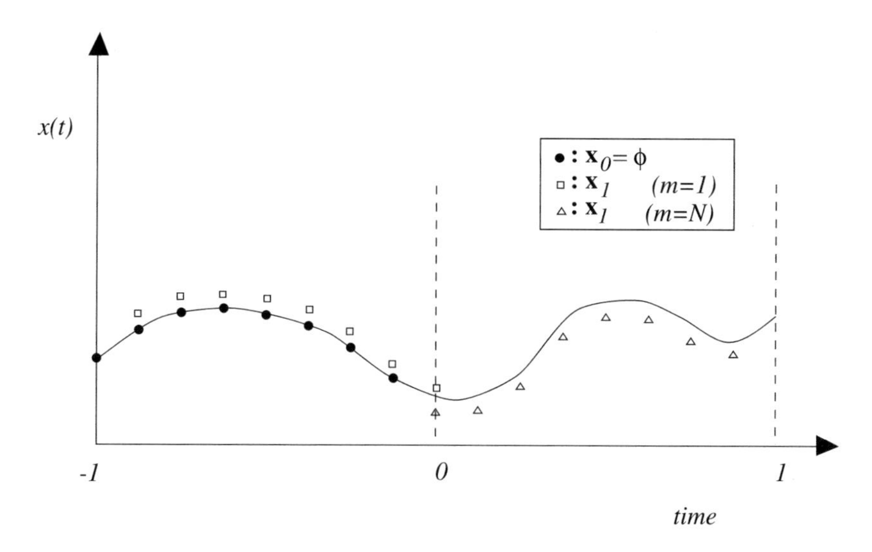

Fig. 7.1 Schematic illustration of the approximation of the differential delay equation (7.1) by a coupled map lattice. The initial function is replaced by a set of N points, and these N points form a vector that evolves in time under the action of an N-dimensional discrete time transformation (the coupled map lattice). The parameter $1 \leq m \leq N$ denotes the number of elements of x_n which are not elements of x_{n+1}. See the text for more detail. Taken from Losson and Mackey (1995) with permission

where $0 < \Delta \ll 1$.

Before describing the second step of the reduction, recall from (7.2) that Eq. (7.1) transforms an initial function ϕ defined on $[-1, 0)$ into another function: the solution x_ϕ defined on $[-1 + t, t)$, where $0 < t \leq 1$ is continuous. Hence, if $t < 1$, there is an overlap between ϕ and x_ϕ. It is possible to vary the extent of this overlap by restricting the values which can be assumed by the time t in the definition (7.2). For example, if $t = m\Delta$, with $0 < \Delta \ll 1$ and $m = 1, 2, \cdots$, the continuous time definition (7.2) can be replaced by

$$S_m\phi = \{x_\phi(s) : s \in [m\Delta - 1, m\Delta)\}, \quad 0 \leq m\Delta < 1. \tag{7.5}$$

If $\Delta = 1/N$, where $N \gg 1$, then $m = 1, \cdots, N$.

The second step in the reduction consists of approximating the initial function ϕ by a set of N points (as illustrated in Fig. 7.1), and following the evolution of these points approximating the corresponding solution.

Hence, if $m = 1$ in (7.5), the initial function ϕ is replaced by a vector $\boldsymbol{\phi} = (\phi_1, \cdots, \phi_N)$, and the solution $\{x_\phi(s) : s \in [\Delta - 1, \Delta)\}$ by a vector $\mathbf{x}_1 = (x_1^1, \cdots, x_1^N)$ (the subscript ϕ has been dropped to simplify the notation). Now define a discrete time transformation $T_1 : \mathbb{R}^N \longmapsto \mathbb{R}^N$ (the subscript indicates that $m = 1$) such that

$$T_1 \circ \overset{n \text{ times}}{\cdots} \circ T_1 \equiv T_1^n(\mathbf{x}_0) = \mathbf{x}_n, \quad n = 1, 2, \cdots, \qquad \text{where } \mathbf{x}_0 \equiv \boldsymbol{\phi}.$$

To obtain an explicit expression for T_1, let $\Delta \equiv 1/N$, and suppose that $\phi_j = \phi(-1 + j\Delta)$, so that in general, x_n^j approximates the value of solution $x(t)$ at time $t = -1 + (n+j)\Delta$. Then, Eq. (7.4) can be approximated by an N-dimensional difference equation

$$\begin{aligned} x_1^1 &= x_0^2 \\ \vdots &= \vdots \\ x_1^{j-1} &= x_0^j \\ \vdots &= \vdots \\ x_1^{N-1} &= x_0^N \\ x_1^N &= \frac{1}{(1+\alpha\Delta)} \left[x_0^N + \Delta S(x_0^1) \right]. \end{aligned} \tag{7.6}$$

In vector notation, the system (7.6) can be written as

$$\mathbf{x}_{n+1} = \mathbf{A}_1 \circ \mathbf{x}_n, \quad \text{for } n = 0, 1, \cdots, \tag{7.7}$$

where the matrix $\mathbf{A}_1$ is given by

$$\mathbf{A}_1 = \begin{pmatrix} 0 & 1 \cdots 0 & 0 \\ 0 & 0 \; 1 \; \cdots & 0 \\ & \ddots & \\ \frac{\Delta S}{(1+\alpha\Delta)} & 0 \cdots 0 & \frac{1}{(1+\alpha\Delta)} \end{pmatrix}. \tag{7.8}$$

Equations (7.7) and (7.8) define a transformation T_1 which approximates the differential delay equation (7.1). In the limit $N \to \infty$, the solution of the difference equation (7.7) converges to the solution of the differential delay equation (7.1), because x is by definition always differentiable. $\mathbf{x}_n$ approximates the continuous time solution on the time interval $[n\Delta - 1, n\Delta)$, and $\mathbf{x}_{n+1}$ approximates the solution on the time interval $[(n+1)\Delta - 1, (n+1)\Delta)$. As illustrated schematically in Fig. 7.1, in general one can approximate the original differential delay equation by a transformation T_m such that $\mathbf{x}_{n+1}$ approximates the solution on $[(n+m)\Delta - 1, (n+m)\Delta)$ (with m an integer such that $1 \le m \le N$, as in (7.5)).

If $m > 1$ in (7.5), the set of difference equations (7.6) becomes

$$
\begin{aligned}
x_1^1 &= x_0^{1+m} \\
\vdots &= \vdots \\
x_1^j &= x_0^{j+m} \\
\vdots &= \vdots \\
x_1^{N-m+1} &= \frac{1}{(1+\alpha\Delta)} \left[x_0^N + \Delta S(x_0^1) \right] \\
\vdots &= \vdots \\
x_1^i &= \frac{1}{(1+\alpha\Delta)} \left[x_1^{i-1} + \Delta S(x_0^{m+1+(N-i)}) \right] \\
\vdots &= \vdots \\
x_1^N &= \frac{1}{(1+\alpha\Delta)} \left[x_1^{N-1} + \Delta S(x_0^{m+1}) \right].
\end{aligned}
$$

Therefore, in vector notation, the equation which generalizes (7.7) is

$$
\mathbf{x}_{n+1} = \mathbf{B}_m \mathbf{x}_{n+1} + \mathbf{A}_m \circ \mathbf{x}_n, \tag{7.9}
$$

where the $N \times N$ matrices $\mathbf{A}_m$ and $\mathbf{B}_m$ are given by

$$
\mathbf{B}_m = \begin{pmatrix}
0 & \cdots & & & & \cdots & 0 \\
\vdots & & [N-(m-1)]_{\text{Empty rows}} & & & \vdots & \\
0 & \cdots & & & & \cdots & 0 \\
0 & \overset{(N-m)\,0's}{\cdots} & 0 & \frac{1}{(1+\alpha\Delta)} & 0 & \cdots & 0 \\
\ddots & & & & & \ddots & \\
0 & \cdots & & 0 & \frac{1}{(1+\alpha\Delta)} & 0 & 0 \\
0 & \cdots & & & 0 & \frac{1}{(1+\alpha\Delta)} & 0
\end{pmatrix} \tag{7.10}
$$

and

$$\mathbf{A}_m = \begin{pmatrix} 0 & \overset{m0's}{\cdots} & & 0 & 1\ 0 \cdots & & \\ 0 & \cdots & & & 0\ 1 & 0 & \cdots \\ \ddots & & & \ddots & & & \\ 0 & \cdots & & & & 0 & 1 \\ \frac{\Delta S}{(1+\alpha\Delta)} & 0 & \cdots & & & 0 & \frac{1}{(1+\alpha\Delta)} \\ 0 & \frac{\Delta S}{(1+\alpha\Delta)} & 0 & \cdots & & & 0 \\ \ddots & & & \ddots & & & \\ 0 & \cdots & 0 & \frac{\Delta S}{(1+\alpha\Delta)} & 0 \cdots & & 0 \end{pmatrix}. \tag{7.11}$$

In the case where $\mathbf{x}_n$ and $\mathbf{x}_{n+1}$ have no overlap, which corresponds to choosing $m = N$, these matrices become:

$$\mathbf{B}_N = \begin{pmatrix} 0 & \cdots & & 0 \\ \frac{1}{(1+\alpha\Delta)} & 0 \cdots & 0 & \\ \ddots & & & \ddots \\ 0 & \cdots & \frac{1}{(1+\alpha\Delta)} & 0 \end{pmatrix}, \quad \mathbf{A}_N = \begin{pmatrix} \frac{\Delta S}{(1+\alpha\Delta)} & 0 & \cdots\ 0 & \frac{1}{(1+\alpha\Delta)} \\ 0 & \frac{\Delta S}{(1+\alpha\Delta)} & 0\ \cdots & 0 \\ \ddots & & \ddots & \\ 0 & \cdots & 0 & \frac{\Delta S}{(1+\alpha\Delta)} \end{pmatrix}.$$

Given the form of the matrix $\mathbf{B}_m$, it is possible to write

$$\mathbf{x}_{n+1} = (\mathbf{I} - \mathbf{B}_m)^{-1}\mathbf{A}_m \circ \mathbf{x}_n \equiv T_m(\mathbf{x}_n).$$

We will assume from now on that S is piecewise linear, because when this is the case, (7.9) can be simplified by replacing composition in the right-hand side by a simple multiplication:

$$\mathbf{x}_{n+1} = (\mathbf{I} - \mathbf{B}_m)^{-1}\mathbf{A}_m\mathbf{x}_n \equiv T_m(\mathbf{x}_n).$$

We have therefore reduced the differential delay equation (7.1) with a piecewise linear S to a piecewise linear problem which can be analyzed from a probabilistic point of view.

This probabilistic analysis is done via an investigation of the spectral properties of the Frobenius–Perron operator associated with T_m.

7.1.1 Functions of Bounded Variation

The transfer operators discussed in this section act on functions which are elements of normed linear spaces. The metric properties of these spaces depend on the choice of the norm. For reasons which will become clear in the next section, two natural norms arise in the descriptions of dynamical systems: the familiar L^1 norm and the so-called *bounded variation* norm. To introduce the latter, it is necessary to recall the definition of the variation of a high-dimensional function. The short discussion given here is based on the presentations of Giusti and Williams (1984), Keller and Künzle (1992), and Ziemer (2012).

First, we define the gradient ∇_{d} in the distributional sense. Let f be a real-valued function defined on an open set $X \subset \mathbb{R}^N$, and $C^1(X)$ denote the space of differentiable functions from X to X having compact support. Then the operator ∇_{d} is the vector valued measure defined by

$$\nabla_{\text{d}} f \equiv \left(\frac{\partial f}{\partial x_1}, \cdots, \frac{\partial f}{\partial x_N} \right).$$

With this definition, it is possible to define the *variation* of f:

$$\bigvee(f) = \|\nabla_{\text{d}} f\|, \tag{7.12}$$

where

$$\|\nabla_{\text{d}} f\| \equiv \sup \left\{ \int_X f \sum_{i=1}^{N} \frac{\partial h^{(i)}}{\partial x_i} d\mu_{\text{L}}^{\mathbb{R}^N} : h = (h^{(1)}, \cdots, h^{(N)}) \in C^1(X), |h| \le 1 \right\},$$

and μ_{L}^X denotes the Lebesgue measure on X. A more detailed presentation is given in Ziemer (2012, Chapter 5) and in Keller and Künzle (1992). Giusti and Williams (1984, Chapter 1) introduce functions of bounded variation in a somewhat more intuitive manner.

With the definition (7.12), it is possible to introduce the bounded variation norm:

$$\|\cdot\|_{BV} \equiv \bigvee(\cdot) + \|\cdot\|_{L^1}. \tag{7.13}$$

The space of functions of bounded variation defined on X is a Banach space (cf. Giusti and Williams 1984) denoted $BV(X)$.

The rationale for introducing the notion of variation, and the associated norm for the probabilistic description of dynamics, is that it allows us to use the following result from the theory of linear operators due to Ionescu Tulcea and Marinescu.

7.1.2 *The Result of Ionescu Tulcea and Marinescu*

This result was originally published in Tulcea and Marinescu (1950) and is of fundamental importance for our analytic description of the probabilistic properties of deterministic systems as examined in this section.

Theorem 7.1 (Tulcea and Marinescu (1950))
Consider two Banach spaces

$$(\mathbb{A}, \| \|_{\mathbb{A}}) \subset (\mathbb{Y}, \| \|_{\mathbb{Y}})$$

with the properties:

1. *If* $\{f_n\}$ *is a bounded sequence of elements of* $\mathbb{A}$ *such that*

$$\lim_{n \longrightarrow \infty} \| f_n - f \|_{\mathbb{Y}} = 0$$

   where $f \in \mathbb{Y}$, *then* f *is also an element of* $\mathbb{A}$, *and* $\| f \|_{\mathbb{A}} \leq \sup_n \| f_n \|_{\mathbb{A}}$.
2. *Let* $P : (\mathbb{A}, \| \|_{\mathbb{A}}) \longmapsto (\mathbb{A}, \| \|_{\mathbb{A}})$ *be a bounded operator which can be extended to a bounded operator in* $(\mathbb{Y}, \| \|_{\mathbb{Y}})$.
3. *Suppose that there is an integer n such that*
   (a) *If* $\mathbb{X}$ *is a* $\| \|_{\mathbb{A}}$*-bounded subset, then* $P^n\mathbb{X}$ *is compact in* $\mathbb{Y}$.
   (b) $\sup_n \| P^n \|_{\mathbb{Y}} < \infty$
   (c) *There exists* $q \in (0, 1)$ *and* $c \geq 0$ *such that*

$$\| P^n f \|_{\mathbb{A}} \leq q \| f \|_{\mathbb{A}} + c \| f \|_{\mathbb{Y}}, \quad \textit{for all } f \in \mathbb{A}.$$

If conditions (1)–(3) are satisfied, then the operator P *is asymptotically periodic and has the spectral decomposition* (2.6).

In order to apply this theorem, we follow Gora and Boyarsky (1989) in choosing $(\mathbb{A}, \| \|_{\mathbb{A}}) = (BV(X), \| \|_{BV})$ included in $(\mathbb{Y}, \| \|_{\mathbb{Y}}) = (L^1(X), \| \|_{L^1})$.

1. Verifying (1) By Giusti and Williams (1984, Theorem 1.9), if $\{f_n\} \in BV(X)$, $\| f_n \|_{BV} \leq K$ for $n = 1, 2 \cdots$ and $f_n \to f$ in L^1, then $f \in BV(X)$ and $\| f \|_{BV} \leq K$.
2. Verifying (2) and (3b): The operators under consideration here are Markov (cf. Sect. 2.1), and their operator norm is 1, hence (2) and (3b) are both verified.
3. Verifying (3a): This property follows from Giusti and Williams (1984, Theorem 1.19).

Hence, the theorem of Ionescu Tulcea and Marinescu guarantees that the transfer operators have the spectral decomposition (2.6) if the condition (3c) is satisfied. By focusing our attention on restrictions of $(BV(X), \| \|_{BV})$ and $(L^1(X), \| \|_{L^1})$ to the corresponding sets of normalized probability densities, and remembering the definition (7.13) of the bounded variation norm, the inequality in (3c) for $n = 1$ becomes:

$$\bigvee (Pf) \leq q \bigvee (f) + \tilde{c}, \qquad \tilde{c} > 0, \tag{7.14}$$

where $\tilde{c} = c + q$. In concrete examples, conditions on the parameters will be obtained such that (7.14) holds, and therefore such that the corresponding Frobenius–Perron operator is asymptotically periodic.

7.2 Applications to Differential Delay Equations

We now derive conditions on the control parameters of deterministic systems which guarantee that the associated Frobenius–Perron operator is asymptotically periodic (i.e., satisfies (2.6)).

7.2.1 *Oscillatory Solutions and Expansion Requirements*

We say that a differential delay equation possesses nontrivial statistical behavior when its solutions are oscillatory and bounded (whether they are periodic, quasi-periodic, or chaotic). Hence, for a given equation, we restrict our attention to the regions of parameter space in which the trajectories are oscillatory. To illustrate this point, we use a model with a piecewise linear transformation S similar to a differential delay equation previously considered by Ershov (1991):

$$S(x) = \begin{cases} ax & \text{if } x < 1/2 \\ a(1-x) & \text{if } x \geq 1/2 \end{cases} \qquad a \in (1, 2]. \tag{7.15}$$

The rationale for choosing this nonlinearity is that the resulting differential delay equation displays a wide array of behaviors which are generic in more general (smooth) systems, while remaining amenable to analytic investigations. In addition, since S maps $[0, 1]$ into itself, we know (Ershov 1991, Section 2.1) that the solutions of the differential delay equation will be bounded if the initial function takes values in $[0, 1]$ and if $a/\alpha \leq 2$.

The first fixed point of Eq. (7.1) with (7.15) is $x_*^{(1)} = 0$. It is locally stable when $a < \alpha$ and unstable when $a > \alpha$. When $a > \alpha$, the equation possesses another fixed point

$$x_*^{(2)} = \frac{a}{a+\alpha},$$

which is linearly stable when

$$2\alpha \geq a > \alpha \quad \text{and} \quad \sqrt{a^2 - \alpha^2} < \cos^{-1}\left(\frac{\alpha}{a}\right).$$

When $\sqrt{a^2 - \alpha^2} = \cos^{-1}\left(\frac{\alpha}{a}\right)$, the fixed point becomes unstable via a Hopf bifurcation, and the solutions of the differential delay equation no longer converge to $x_*^{(2)}$. As mentioned above, the solutions must remain bounded when the initial function belongs to the interval $[0, 1]$, and since they do not converge to the fixed point, they must oscillate. We restrict our discussion of the dynamics of (7.1) to regions of parameter space in which the solutions are oscillatory, because stationary solutions are trivial from a statistical perspective. Hence our description of the probabilistic properties of (7.1) with (7.15) holds when the parameters of the equation satisfy

$$2\alpha \geq a > \alpha \quad \text{and} \quad \sqrt{a^2 - \alpha^2} > \cos^{-1}\left(\frac{\alpha}{a}\right). \tag{7.16}$$

When condition (7.16) is satisfied, the corresponding map T_m expands distances in at least one direction. To see this, note that from (7.9) with (7.10) and (7.11), the total derivative of the variable x_{n+1}^{N-m} is given by

$$\begin{aligned}\frac{dx_{n+1}^{N-m}}{d\mathbf{x}_n} &= \frac{\partial x_{n+1}^{N-m}}{\partial x_n^N} + \frac{\partial x_{n+1}^{N-m}}{\partial x_n^1} \\ &= \frac{N+a}{N+\alpha},\end{aligned}$$

and so $\frac{dx_{n+1}^{N-m}}{d\mathbf{x}_n} > 1$ if and only if $a > \alpha$. If condition (7.16) is satisfied, this is always the case, and therefore when the differential delay equation possesses oscillatory solutions, the corresponding map is hyperbolic (or expanding if one chooses to define expansion by the requirement that at least one eigendirection be expanding).

Examples of oscillatory solutions of Eq. (7.1) with S given by (7.15) are shown in Fig. 7.2. The parameters used to produce that figure are the same as the ones used to produce the "ensemble density" results presented in Fig. 3.6. As expected, the remarkable agreement between the solutions obtained by both methods breaks down when N becomes too small (i.e., of order 10^2), and for large times when the solution is chaotic.

Having derived the approximation to the differential delay equation, we now use this expression to rigorously discuss the probabilistic behavior of the equation.

7.2.2 The Result

When the parameters α and a of the differential delay equation (7.1), with nonlinearity S given by (7.15), satisfy (7.16), and the initial function ϕ for the equation belongs to the interval [0, 1], the corresponding approximation (7.9) induces a Frobenius–Perron operator which is asymptotically periodic and therefore has the spectral decomposition (2.6).

We now prove this statement.

Using basic properties of determinants (Muir and Metzler 2003), for all $\mathbf{x} \in X$, and all $i = 1, \cdots, M$ the Jacobian of $T^{-1}(\mathbf{x})$ is

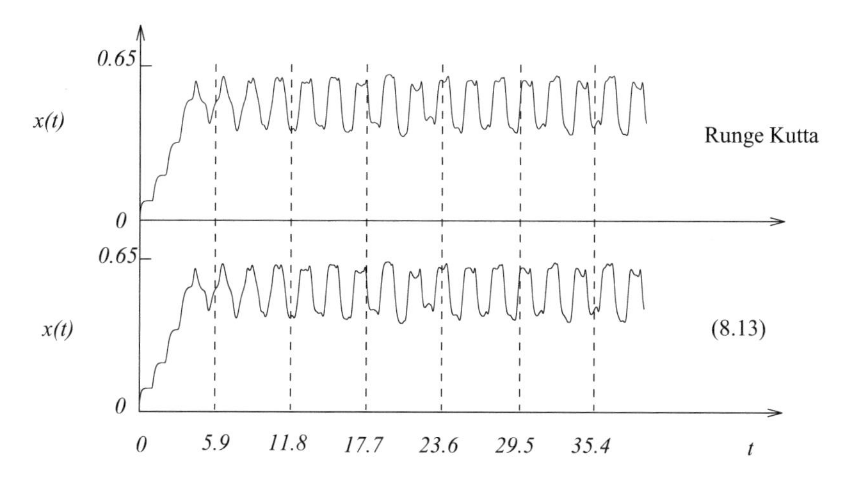

Fig. 7.2 Two numerical solutions of the differential delay equation (7.1) with the nonlinearity given by (7.15), when $a = 13, \alpha = 10$, and a constant initial function $\phi(s) = 0.2$ for $s \in [-1, 0]$. Top: The solution was produced by a standard adaptation of the fourth order Runge–Kutta method, with 40 points per delay. Bottom: The solution was produced by the approximation (7.9), with $m = N = 1000$. As expected, although both solutions are in excellent agreement with one another, the Runge–Kutta method is numerically more efficient than the Euler approximation which underlies our derivation of the approximation. Taken from Losson and Mackey (1995) with permission

$$J(T_{|i}^{-1}(\mathbf{x})) = J^{-1} = \frac{\det\,(\mathbf{I} - |\mathbf{B}'_m|)}{\det\,|\mathbf{A}'_m|},$$

where $|\mathbf{A}'_m|$ and $|\mathbf{B}'_m|$ can be obtained by replacing S by a in $\mathbf{A}_m$ (defined in (7.11)) and $\mathbf{B}_m$ (defined in (7.10)) respectively. It is straightforward to verify that

$$J^{-1} = \left[\frac{(N+\alpha)}{a}\right]^m. \tag{7.17}$$

If the transformation T is piecewise invertible, it is possible to give a more explicit definition of P_T. Let $T_{|i}$ be the invertible restriction of T to the set $\pi_i \subset X$, $i = 1, \cdots, M$ (with of course $\bigcup_{i=1}^{M} \pi_i = X$). Let $\tilde{\pi}_i$ denote the image of the set π_i: $\tilde{\pi}_i \equiv T_{|i}(\pi_i)$. The Frobenius–Perron operator induced by T can be written as

$$f_{n+1}(\mathbf{x}) \equiv P_T f_n(\mathbf{x}) = \sum_{i=1}^{M} \frac{f_n(T_{|i}^{-1}(\mathbf{x}))}{J(T_{|i}^{-1}(\mathbf{x}))} \chi_{\tilde{\pi}_i}(\mathbf{x}), \tag{7.18}$$

where $\chi_{\tilde{\pi}_i}(\mathbf{x}) \equiv 1$ iff $\mathbf{x} \in \tilde{\pi}_i$, and 0 otherwise, and $J(Z)$ is the absolute value of the Jacobian of Z. A more intuitive version of (7.18) is

$$P_T f_n(\mathbf{x}) = \sum_{\mathbf{y} \in T^{-1}(\mathbf{x})} \frac{f_n(\mathbf{y})}{J(T(\mathbf{y}))}.$$

Therefore, using the definition (7.18), basic properties of the variation and (7.17)

$$\begin{aligned}
\bigvee(P_T f) &= \bigvee\left(\sum_{i=1}^{M} \frac{f(T_{|_i}^{-1}(\mathbf{x}))}{J(T_{|_i}^{-1}(\mathbf{x}))} \chi_{\tilde{\pi}_i}(\mathbf{x})\right) \\
&\leq \sum_{i=1}^{M} \bigvee\left(\frac{f(T_{|_i}^{-1}(\mathbf{x}))}{J(T_{|_i}^{-1}(\mathbf{x}))} \chi_{\tilde{\pi}_i}(\mathbf{x})\right) \\
&= J \sum_{i=1}^{M} \bigvee\left(f(T_{|_i}^{-1}(\mathbf{x}))\chi_{\tilde{\pi}_i}(\mathbf{x})\right).
\end{aligned} \tag{7.19}$$

Each term in the sum on the right-hand side of (7.19) can now be evaluated explicitly. From the definition (7.12),

$$\begin{aligned}
&\bigvee\left(f(T_{|_i}^{-1}(\mathbf{x}))\chi_{\tilde{\pi}_i}(\mathbf{x})\right) = \left\|\nabla_{\mathrm{d}}\left[f(T_{|_i}^{-1}(\mathbf{x}))\chi_{\tilde{\pi}_i}(\mathbf{x})\right]\right\|_{L^1} \\
&= \int_X |\nabla_{\mathrm{d}}\left[f(T_{|_i}^{-1}(\mathbf{x}))\right] \chi_{\tilde{\pi}_i}(\mathbf{x})|\, d\mu_{\mathrm{L}}^X + \int_{\mathbb{R}^N} |f(T_{|_i}^{-1}(\mathbf{x}))\nabla_{\mathrm{d}}\left[\chi_{\tilde{\pi}_i}(\mathbf{x})\right]|\, d\mu_{\mathrm{L}}^X \\
&= \bigvee\left(f(T_{|_i}^{-1}(\mathbf{x}))\right)\Big|_{\mathbf{x}\in\tilde{\pi}_i} + \int_{\mathbb{R}^N} |f(T_{|_i}^{-1}(\mathbf{x}))\nabla_{\mathrm{d}}\left[\chi_{\tilde{\pi}_i}(\mathbf{x})\right]|\, d\mu_{\mathrm{L}}^{\mathbb{R}^N}.
\end{aligned} \tag{7.20}$$

Since J is independent of $\mathbf{x}$, a simple change of variables yields

$$\bigvee\left(f(T_{|\pi_i}^{-1}(\mathbf{x}))\right)\Big|_{\mathbf{x}\in\tilde{\pi}_i} = J \bigvee(f(\mathbf{x}))\Big|_{\mathbf{x}\in\pi_i}. \tag{7.21}$$

The integral on the RHS of Eq. (7.20) can be simplified using Giusti and Williams (1984, Example 1.4), which states that for any $g \in BV(X)$, and $B \subset X$ with piecewise C^2 boundaries of finite $N-1$-dimensional measure,

$$\int_X |g(\mathbf{x})\nabla_{\mathrm{d}}[\chi_B(\mathbf{x})]|\, d\mu_{\mathrm{L}}^{\mathbb{R}^N} = \int_{\partial B} |g(\mathbf{x})|\, d\mu_{\mathrm{L}}^{\mathbb{R}^{N-1}}.$$

Choosing $g(\mathbf{x}) = f(T_{|_i}^{-1}(\mathbf{x}))$, and $B = \tilde{\pi}_i$, one obtains

$$\int_X |f(T_{|_i}^{-1}(\mathbf{x}))\nabla_{\mathrm{d}}[\chi_{\tilde{\pi}_i}(\mathbf{x})]\, d\mu_{\mathrm{L}}^{\mathbb{R}^N} = \int_{\partial\tilde{\pi}_i} |f(T_{|_i}^{-1}(\mathbf{x}))|\, d\mu_{\mathrm{L}}^{\mathbb{R}^{N-1}}.$$

Furthermore, for any $g \in BV(X)$, and any B as specified above, we have from Gora and Boyarsky (1989, Lemma 3)

$$\int_{\partial B} g(\mathbf{x}) d\mu_{\mathrm{L}}^{\mathbb{R}^{N-1}} \leq \frac{1}{\sin\theta(B)} \bigvee (g(\mathbf{x}))\Big|_{\mathbf{x}\in B} + K_B,$$

where $K_B > 0$ is bounded and $\sin\theta(B)$ depends on the smallest angle subtended by intersecting edges of the set B. Letting $g(\mathbf{x}) = f(T_{|_i}^{-1}(\mathbf{x}))$ and $B = \tilde{\pi}_i$ as before, and recalling identity (7.21), the integral in (7.20) satisfies

$$\begin{aligned}\int_X |f(T_{|_i}^{-1}(\mathbf{x}))\nabla_{\mathrm{d}}\left[\chi_{\tilde{\pi}_i}(\mathbf{x})\right]| \, d\mu_{\mathrm{L}}^{\mathbb{R}^N} &\leq \frac{1}{\sin\theta(\tilde{\pi}_i)} \bigvee \left(f(T_{|_i}^{-1}(\mathbf{x}))\right)\Big|_{\mathbf{x}\in\tilde{\pi}_i} + K_i \\ &= \frac{J}{\sin\theta(\tilde{\pi}_i)} \bigvee (f(\mathbf{x}))\Big|_{\mathbf{x}\in\pi_i} + K_i. \qquad (7.22)\end{aligned}$$

Letting $\sin\theta(\tilde{\pi}) \equiv \min_i \sin\theta(\tilde{\pi}_i)$, and using (7.20), (7.21), and (7.22), (7.19) becomes

$$\begin{aligned}\bigvee(P_T f) &\leq J \sum_{i=1}^{M} \left[J \bigvee (f(\mathbf{x}))\Big|_{\mathbf{x}\in\pi_i} + \frac{J}{\sin\theta(\tilde{\pi}_i)} \bigvee (f(\mathbf{x}))\Big|_{\mathbf{x}\in\pi_i} + K_i \right] \\ &\leq J^2 \bigvee (f) \left[1 + \frac{1}{\sin\theta(\tilde{\pi})}\right] + M \max_i K_i. \qquad (7.23)\end{aligned}$$

Therefore, comparing (7.23) (using (7.17)) with (7.14), the theorem of Ionescu Tulcea and Marinescu guarantees the asymptotic periodicity of the Frobenius–Perron operator when

$$\left[\frac{a}{(N+\alpha)}\right]^{2m} \left[1 + \frac{1}{\sin\theta(\tilde{\pi})}\right] < 1. \qquad (7.24)$$

In order for this inequality to yield precise conditions on the parameters, it is necessary to evaluate the quantity $\sin\theta(\tilde{\pi})$ in terms of these parameters. This calculation is straightforward but lengthy and has been published for general situations of which the present problem is a special case (Losson et al. 1995). If the boundaries of the sets $\tilde{\pi}_i$ intersect at an angle which is bounded below by $\omega > 0$ (for all $i = 1, \cdots, M$), then when X, and thus the π_i's, are subsets of $\mathbb{R}^N$, we have (Losson et al. 1995)

$$\sin\theta(\tilde{\pi}) = \sqrt{\frac{1 - \cos\omega}{N[1 + (N-2)\cos\omega]}}. \qquad (7.25)$$

Note that if the boundaries of the image sets intersect at right angles so that $\omega = \pi/2$, we have $\sin\theta(\tilde{\pi}) = 1/\sqrt{N}$. In general, however, the image partition is not rectangular, and the angle ω must be determined from the definition of the map under consideration.

Recall that the sets π_i are defined to be the subsets of X on which the map given in (7.9) is piecewise linear. If S is given by (7.15), these sets are of the form $\pi_i = [0, 1/2]^i \times (1/2, 1]^{2^N - i}$, since the solution $x(t)$ of the original differential delay equation (7.1) satisfies $0 \le x(t) \le 1$ for all $t > 0$ by hypothesis. Hence the π_i's are delimited by 3^N vertices, $\{\mathbf{v}_l = (v_l^{(1)}, \cdots, v_l^{(N)})\}$, $l = 1, \cdots, 3^N$ where each component $v_l^{(k)}$ is either 0, 1/2, or 1 for $k = 1, \cdots, N$. The images $\tilde{\mathbf{v}}_l$ of the $\mathbf{v}_l$'s are the vertices of the image partition whose elements are denoted $\tilde{\pi}_i$.

From Eqs. (7.9)–(7.11), we have

$$
\begin{aligned}
\tilde{v}_l^{(k)} &= v_l^{(k+m)}, \quad \text{for } k = 1, \cdots, N - m \\
\tilde{v}_l^{(N-m+1)} &= \frac{\Delta S(v_l^{(1)})}{(1+\alpha\Delta)} + \frac{v_l^{(N)}}{(1+\alpha\Delta)} \\
\tilde{v}_l^{(N-m+j)} &= \sum_{p=1}^{j} (-1)^{(p-1)} \left[\frac{\Delta S(v_l^{(j+1-p)})}{(1+\alpha\Delta)} \right]^p \quad \text{for } j = 2, \cdots, m.
\end{aligned}
\tag{7.26}
$$

Determining the smallest angle ω which can be subtended by edges of one of the image sets $\tilde{\pi}_i$ is a tedious minimization problem. However, a conservative lower bound for ω can be obtained relatively easily. If this lower bound is used in (7.25), we obtain a lower bound for $\sin\theta(\tilde{\pi})$, and the resulting conditions on a, α, m and N obtained from (7.24) will be correspondingly conservative.

The angle ω is defined by three points which are images of three of the 3^N vertices delimiting the sets π_i. The lower bound for ω is defined by a triplet such that two of the three elements are distinct but as close as (7.26) permits, while being as far from the third as possible. Hence, to determine the lower bound for ω, two distances must be estimated. The first is the greatest Euclidean distance separating two points belonging to the same face of one of the image sets (which is bounded above by $\sqrt{N}$, the "diameter" of the phase space X). The second is the smallest distance $d_{\tilde{l}\tilde{l}'}$ separating two images $\tilde{\mathbf{v}}_l$ and $\tilde{\mathbf{v}}_{l'}$ ($l \neq l'$),

$$
d_{\tilde{l}\tilde{l}'} = \sqrt{\sum_{k=1}^{N} [\tilde{v}_l^{(k)} - \tilde{v}_{l'}^{(k)}]^2}.
$$

Minimizing this distance is straightforward since all the terms are positive, so we must first pick $\mathbf{v}_l$ and $\mathbf{v}_{l'}$ such that their images differ in only one of their components, and then try to minimize the resulting difference by choosing this component appropriately.

There are two ways to choose $\mathbf{v}_l$ and $\mathbf{v}_{l'}$ such that $\tilde{\mathbf{v}}_l$ and $\tilde{\mathbf{v}}_{l'}$ differ in a single component.

1. From (7.26), if $\mathbf{v}_l$ and $\mathbf{v}_{l'}$ differ by only one component which belongs to both the preimage $\mathbf{v}_{l,l'}$ and the image $\tilde{\mathbf{v}}_{l,l'}$ (i.e., $v_l^{(i)} = v_{l'}^{(i)}$ for all i except for $i = k$,

with $m < k < N$), then $d_{\tilde{l}\tilde{l}'}$ must equal the distance separating $\mathbf{v}_l$ and $\mathbf{v}_{l'}$. This nonzero distance is bounded below by $1/2$.

2. It is possible to obtain a more conservative (i.e., smaller) estimate for $d_{\tilde{l}\tilde{l}'}$ by noting that in the regions of parameter space which are of interest, from (7.26), T_m is a contraction along some directions. If $\mathbf{v}_l$ and $\mathbf{v}_{l'}$ differ in their k^{th} component with $1 \leq k \leq m$, then their images $\tilde{\mathbf{v}}_l$ and $\tilde{\mathbf{v}}_{l'}$ will possess $(k - m + 1)$ different components $\{j = N - (k - m + 1), N - (k - m) + 1, \cdots, N\}$. By choosing $k = m$, $\tilde{\mathbf{v}}_l$ and $\tilde{\mathbf{v}}_{l'}$ differ in a single component.

In the second case, the distance between $\tilde{\mathbf{v}}_l$ and $\tilde{\mathbf{v}}_{l'}$ is minimized, and from (7.26) it is easily shown to be

$$\begin{aligned} d_{\tilde{l}\tilde{l}'} = |\tilde{v}_l^{(N)} - \tilde{v}_{l'}^{(N)}| &= \left| \frac{S(v_l^m) - S(v_{l'}^m)}{(N + \alpha)} \right| \\ &\geq \frac{a}{2(N + \alpha)}, \end{aligned}$$

since if $S(v_l^m) - S(v_{l'}^m) \neq 0$, then $S(v_l^m) - S(v_{l'}^m) \geq a/2$.

From plane analytic geometry, ω is therefore bounded below by the angle subtended by a horizontal line, and a line of slope $\frac{a}{2\sqrt{N}(N+\alpha)}$, so

$$\tan \omega = \frac{a}{2\sqrt{N}(N + \alpha)}.$$

Therefore, if N is large we have, using the approximations $\cos \omega \simeq 1 - \omega$, and $\sin \omega \simeq \omega$ (for small ω)

$$\omega \simeq \frac{a}{2\sqrt{N}(N + \alpha)}.$$

Using the same approximations in (7.25), we obtain

$$\sin \theta(\tilde{\pi}) \simeq \sqrt{\frac{a}{2N^{3/2}(N + \alpha)[1 + (N - 2)]}}.$$

Replacing this conservative estimate in the condition (7.24), we finally obtain an explicit condition on the parameters of the map which is sufficient to guarantee asymptotic periodicity:

$$\left[\frac{a}{(N + \alpha)}\right]^{2m} \left[1 + \sqrt{\frac{2N^{3/2}(N + \alpha)[1 + (N - 2)]}{a}}\right] < 1. \quad (7.27)$$

If N is large, the left-hand side of the preceding inequality behaves like $N^{-2m} N^{7/4}$. Hence the inequality is always satisfied for N large enough, and Theorem 7.1

implies that P_{T_m} is always asymptotically periodic, when T_m approximates the differential delay equation (7.1) with S given by (7.15) under the condition (7.16) (though the period could be 1).

Numerically, this result is reflected by the temporally periodic behavior of various statistical descriptors of the motion. As an example, consider the "ensemble sample density" $f(x, t)$. This function is obtained by integrating (7.1) with a large number of different initial functions $\{\phi_1, \cdots, \phi_E\}$ (E large) and then, at time t, binning the set of points $\{x_{\phi_i}(t)\}$, where $x_{\phi_i}(t)$ denotes the solution of the differential delay equation corresponding to the initial function ϕ_i. Schematically, Fig. 7.1 displays this construction. To establish a parallel with more frequently discussed models, if the equation satisfied by $x(t)$ was an ordinary differential equation (rather than a differential delay equation), the evolution of $f(x, t)$ would be described by the Liouville equation.

The statistical cycling predicted by (7.27) can be observed numerically by following the function $f(x, t)$ for successive times. Figure 3.6 displays such a numerical simulation for the differential delay equation (7.1) with S defined in (7.15). The novel feature displayed in Fig. 3.6 is the dependence of the asymptotic density cycle on the initial density which describes the set of initial functions used to carry out a set of simulations. This property is not observed in continuous-time systems without delays, and it can be understood in light of the dependence of the functionals $\lambda_{1,\cdots,r}$ of Eq. (2.6) on the initial density f_0. This dependence on initial conditions is in a sense much stronger than that usually discussed in relation to chaotic dynamical systems: Here the evolution of an ensemble of differential delay equations depends on the exact distribution of the initial ensemble.

7.3 Summary and Conclusions

This chapter considers approximations in which we turn a differential delay equation into a high-dimensional map through a process of discretization. Although they offer useful approximations, it is only in Sect. 7.2 that we are able to present analytic results based on the work of Tulcea and Marinescu (1950), and even then it is unclear how to extend these results to the limiting case where we recover the differential delay equation from the discretized map. However, this may be a promising avenue for further exploration.

Chapter 8
Approximate “Liouville-Like” Equation and Invariant Densities for Delay Differential Equations

This entire chapter is taken from an unpublished manuscript (Taylor 2011).

8.1 Discretized Frobenius–Perron Operator

Consider the following delay differential equation with nonlinear delayed feedback,

$$x'(t) = -x(t) + S(x(t - \tau)). \tag{8.1}$$

Equations of this form occur frequently in diverse applications, and their solutions give rise to a variety of dynamical forms (an der Heiden and Mackey 1982). We wish to determine the evolution of the probability density associated with an ensemble of solutions of (8.1), arising, e.g., from an ensemble of initial conditions, or from uncertain initial conditions with a prescribed probability distribution. Such a probabilistic approach to delay equations has been pursued by other researchers (Losson and Mackey 1992, 1995; Ershov 1991; Capiński 1991; an der Heiden 1985; Dorrizzi et al. 1987; Lepri et al. 1993). Here we proceed by developing the Frobenius–Perron operator (Lasota and Mackey 1994) corresponding to a discrete-time map derived from a discretization of Eq. (8.1).

With $y(t) = x(t - \tau)$ Eq. (8.1) can be written as

$$x'(t) = -x(t) + S(y(t)). \tag{8.2}$$

Discretizing in t with step size h yields the Euler time step

$$\begin{aligned} x_{n+1} &= x_n + h(-x_n + S(y_n)) \\ &= (1 - h)x_n + hS(y_n) \\ &\equiv T(x_n, y_n). \end{aligned} \tag{8.3}$$

J. Losson et al., *Density Evolution Under Delayed Dynamics*, Fields Institute Monographs 38, https://doi.org/10.1007/978-1-0716-1072-5_8

Consider an ensemble of solutions of (8.2), and denote by $\rho_n(x, y)$ the bivariate density at time step n of the corresponding ensemble of $(x_n, y_n) \in \mathbb{R}^2$. Then, at the following time step, the ensemble of $x_{n+1} = T(x_n, y_n)$ has some density which we denote $f_{n+1}(x)$. The Frobenius–Perron operator corresponding to T is just the operator $P\colon L^1(\mathbb{R}^2) \to L^1(\mathbb{R})$ such that $P\rho_n = f_{n+1}$. In other words, P carries densities forward in time under the action of T. (Note that one cannot determine the evolution of the density $f(x)$ without introducing the bivariate density $\rho(x, y)$, since the evolution of x depends directly via Eq. (8.1) on the delayed coordinate y. In much the same way, an evolution equation for ρ will involve a trivariate density depending on an additional delayed coordinate, and so on.)

The operator P is uniquely determined by the following relation (Lasota and Mackey 1994):

$$\int_A (P\rho_n)(x)\, dm(x) = \int_{T^{-1}(A)} \rho_n(x, y)\, dm^2(x, y), \tag{8.4}$$

where $A \subset \mathbb{R}$ is an arbitrary measurable set, and m, m^2 denote Lebesgue measure on $\mathbb{R}$ and $\mathbb{R}^2$, respectively. Letting $A = [a, x]$ and differentiating (8.4) yield

$$(P\rho_n)(x) = \frac{d}{dx} \int_{T^{-1}[a,x]} \rho_n(r, s)\, dr\, ds. \tag{8.5}$$

Equation (8.3) can be used to find $T^{-1}[a, x]$ explicitly, since

$$\begin{aligned} a < T(r, s) < x &\implies a < (1-h)r + hS(s) < x \\ &\implies \frac{a - hS(s)}{1-h} < r < \frac{x - hS(s)}{1-h}, \end{aligned}$$

hence

$$T^{-1}[a, x] = \left\{ (r, s) : r \in \left[\frac{a - hS(s)}{1-h}, \frac{x - hS(s)}{1-h} \right],\ s \in \mathbb{R} \right\},$$

and Eq. (8.5) becomes

$$\begin{aligned} (P\rho_n)(x) &= \frac{d}{dx} \int_{-\infty}^{\infty} \int_{(a-hS(s))/(1-h)}^{(x-hS(s))/(1-h)} \rho_n(r, s)\, dr\, ds \\ &= \frac{1}{1-h} \int_{-\infty}^{\infty} \rho_n\left(\frac{x - hS(y)}{1-h}, y \right) dy. \end{aligned} \tag{8.6}$$

Here we have an explicit formula for the Frobenius–Perron operator corresponding to the discretized evolution map T of Eq. (8.3).

8.2 Liouville-Like Equation

Letting $P(\rho_n) = f_{n+1}$ and linearizing Eq. (8.6) about $h = 0$ (assuming sufficient differentiability of ρ_n) we obtain

$$
\begin{aligned}
(1-h) f_{n+1}(x) &= \int_{-\infty}^{\infty} \left[\rho_n(x, y) + h\big(x - S(y)\big)\frac{\partial \rho_n}{\partial x}(x, y) + O(h^2) \right] ds \\
&= f_n(x) + \int_{-\infty}^{\infty} \left[h\big(x - S(y)\big)\frac{\partial \rho_n}{\partial x}(x, y) + O(h^2) \right] dy,
\end{aligned}
$$

where we have used the fact that

$$
\int_{-\infty}^{\infty} \rho_n(x, y)\, dy = f_n(x),
$$

i.e., the bivariate density $\rho_n(x, y)$ must project to the univariate marginal density $f_n(x)$, since both describe the same ensemble. On rearrangement this yields

$$
\frac{f_{n+1}(x) - f_n(x)}{h} = f_{n+1}(x) + \int_{-\infty}^{\infty} \left[\big(x - S(y)\big)\frac{\partial \rho_n}{\partial x}(x, y) + O(h) \right] dy,
$$

which reduces in the limit $h \to 0$ to

$$
\begin{aligned}
\frac{\partial f(x,t)}{\partial t} &= f(x,t) + \int_{-\infty}^{\infty} \big(x - S(y)\big)\frac{\partial \rho}{\partial x}(x, y, t)\, dy \\
&= f(x,t) + x\frac{\partial}{\partial x}\int_{-\infty}^{\infty} \rho(x, y, t)\, dy - \frac{\partial}{\partial x}\int_{-\infty}^{\infty} S(y)\rho(x, y, t)\, dy \\
&= f(x,t) + x\frac{\partial f(x,t)}{\partial x} - \frac{\partial}{\partial x}\int_{-\infty}^{\infty} S(y)\rho(x, y, t)\, dy.
\end{aligned}
$$

This last result can be rewritten as

$$
\frac{\partial f}{\partial t} = -\frac{\partial}{\partial x}\left(-x f(x) + \int_{-\infty}^{\infty} S(y)\rho(x, y)\, dy \right). \tag{8.7}
$$

Equation (8.7) is analogous to the Liouville equation

$$
\frac{\partial f}{\partial t} = -\frac{\partial}{\partial x}(F f)
$$

for the evolution of the density for the ordinary differential equation

$$
\frac{dx}{dt} = F(x).
$$

Indeed, we can interpret the differential delay equation (8.2) as an *ordinary* differential equation

$$\frac{dx}{dt} = -x + S(y) = F(x; y)$$

(i.e., as in the method of steps (Diekmann et al. 1995; Hale and Lunel 2013)) in which y is a distributed parameter whose distribution depends on x. Equation (8.7) is just the corresponding Liouville equation, in which the term

$$\int_{-\infty}^{\infty} S(y)\rho(x, y)\, dy$$

is just the conditional expectation of the forcing term $S(y)$ for given x.

8.3 Invariant Densities

Equation (8.7) implies that an invariant density $f(x)$ (together with the invariant bivariate density $\rho(x, y)$) must satisfy

$$\frac{\partial}{\partial x}\left(-xf(x) + \int_{-\infty}^{\infty} S(y)\rho(x, y)\, dy\right) = 0,$$

and hence

$$xf(x) - \int_{-\infty}^{\infty} S(y)\rho(x, y)\, dy = C$$

for some constant C. With $C \neq 0$ this results in

$$f(x) = Cx^{-1} + x^{-1}\int_{-\infty}^{\infty} S(y)\rho(x, y)\, dy,$$

which gives a non-integrable singularity; we take $C = 0$ to eliminate this singularity and obtain

$$xf(x) = \int_{-\infty}^{\infty} S(y)\rho(x, y)\, dy. \tag{8.8}$$

Since $x(t)$ and $y(t) = x(t - \tau)$ represent shifted versions of the same time series, they must share the same invariant density. Therefore, both marginals of ρ must be consistent with f:

$$f(x) = \int_{-\infty}^{\infty} \rho(x, y)\, dy = \int_{-\infty}^{\infty} \rho(y, x)\, dy. \tag{8.9}$$

One interesting consequence is that integrating Eq. (8.8) gives

$$\langle x \rangle = \langle S(x) \rangle, \tag{8.10}$$

where $\langle \cdot \rangle$ denotes expectation with respect to the invariant distribution.

8.4 Copulas

Building the constraint (8.9) into Eq. (8.8) presents a challenge. For any given univariate density f, there is an (uncountably) infinite set of bivariate densities ρ whose marginal distributions both agree with f as in (8.9). Clearly Eqs. (8.8) and (8.9) are underdetermined.

There are several ways to construct a bivariate distribution having given marginals. Any one such method might be used to uniquely determine ρ (somewhat arbitrarily) in terms of f, thus enforcing Eq. (8.9) and providing closure to Eqs. (8.8) and (8.9), which we can then hope to solve to obtain an approximate invariant density f.

For example, if it is known (or assumed) that x, y are independently and identically distributed, each with density f, then their bivariate density $\rho(x, y)$ is given by

$$\rho(x, y) = f(x) f(y).$$

Under this condition Eq. (8.8) becomes

$$x f(x) = f(x) \int_{-\infty}^{\infty} S(y) f(y)\, dy \implies f(x)\big(x - \langle f \rangle\big) = 0.$$

This implies that $x = \langle f \rangle$ (a constant) everywhere within the support of f, a contradiction that seems to provide no useful information.

A more general approach to parametrizing families of bivariate densities ρ that satisfy (8.9) is to use *copulas* (Genest and Mackay 1986; Nelsen 2006). Briefly, a copula is a bivariate distribution function $C(x, y)$ on the unit square with uniform marginals. By Sklar's theorem (Sklar 1959), for any bivariate distribution function $H(x, y)$ with continuous marginal distributions $F(x)$ and $G(y)$, there corresponds a unique copula C such that

$$H(x, y) = C\big(F(x), G(y)\big). \tag{8.11}$$

Conversely, if a copula C and distributions F, G are given, then Eq. (8.11) shows how to construct a bivariate distribution that has marginals F, G. The role of the copula is to specify the dependence structure that binds the marginals of a bivariate distribution. Differentiating Eq. (8.11) gives the bivariate density

$$h(x, y) = c\big(F(x), G(y)\big) f(x)g(y), \tag{8.12}$$

where f, g are the densities of the marginals, and c is the copula density. (For example, the case of independently distributed x, y corresponds to the copula density $c = 1$.)

Let R be the distribution function corresponding to the density f:

$$R(x) = \int_{-\infty}^{x} f(s)\, ds.$$

Then by Eq. (8.12), for any given copula density c the bivariate density

$$\rho(x, y) = c\big(R(x), R(y)\big) f(x) f(y) \tag{8.13}$$

will have marginals that agree with f, thus satisfying Eq. (8.9). Substituting (8.13) in Eq. (8.8) yields

$$xf(x) = f(x) \int_{-\infty}^{\infty} S(y)c\big(R(x), R(y)\big) f(y)\, dy,$$

and thus

$$f(x)\left[x - \int_{-\infty}^{\infty} S(y)c\big(R(x), R(y)\big) f(y)\, dy\right] = 0. \tag{8.14}$$

Consequently, at every x in its support f satisfies the integral equation

$$x = \int_{-\infty}^{\infty} S(y)c\big(R(x), R(y)\big) f(y)\, dy,$$

a nonlinear Fredholm equation of the first kind (Press et al. 1992). Unfortunately, without a priori knowledge of the support of f, this integral equation does not appear to be useful.

8.5 Approximate Invariant Densities

With the aim of discretizing Eq. (8.14), let x_i $(i = 1, \ldots, n)$ be a uniform grid such that the interval $[x_1 = a, x_n = b]$ contains the support of f, and let w_i be quadrature weights appropriate to the x_i so that

$$\int_a^b g(x)\,dx \approx \sum_{i=1}^n w_i g(x_i)$$

for a suitable class of functions g. Evaluating the integral in Eq. (8.14) by numerical quadrature and collocating at the x_i give

$$f_i\left[x_i - \sum_{j=1}^n c\big(R(x_i), R(x_j)\big) w_j f_j s_j\right] = 0, \tag{8.15}$$

where $f_i = f(x_i)$, $s_i = S(x_i)$ $(i = 1, \ldots, n)$. Let r_{ij} be a matrix of quadrature weights so that

$$R(x_j) = \int_a^{x_j} f(s)\,ds \approx \sum_{k=1}^n r_{jk} f_k.$$

Then Eq. (8.15) can be approximated by

$$f_i\Big[x_i - \sum_{j=1}^n c\Big(\sum_{k=1}^n r_{ik} f_k, \sum_{k=1}^n r_{jk} f_k\Big) w_j f_j s_j\Big] = 0 \quad (i = 1, \ldots, n). \tag{8.16}$$

This gives a system of n nonlinear algebraic equations for the unknowns f_i. To eliminate the trivial solution, we augment equations (8.16) with the normalization constraint

$$\int_a^b f(s)\,ds \approx \sum_{i=1}^n w_i f_i = 1. \tag{8.17}$$

Discretization of a Fredholm integral equation typically leads to an ill-conditioned (and over-determined) system of equations, a difficulty that can be circumvented by a variety of regularization methods. Since the unknown invariant density is determined by a Fredholm equation, we expect the same difficulty to arise here.

Denoting the left-hand side of Eq. (8.16) by $\Theta(\boldsymbol{f})$, we regularize by replacing Eqs. (8.16) and (8.17) with the constrained optimization problem

$$\begin{aligned} \min_{\boldsymbol{f}} : \quad & |\Theta(\boldsymbol{f})|^2 + \lambda \boldsymbol{f}^{\mathrm{T}} H \boldsymbol{f} \\ \text{subject to:} \quad & \boldsymbol{f} \cdot \mathbf{w} = 1, \end{aligned} \tag{8.18}$$

where $\lambda \geq 0$ is a real parameter and $H = B^{\mathrm{T}}B$, where B is a finite difference matrix, e.g.,

$$B = \begin{bmatrix} -1 & 1 & 0 & 0 & \cdots & 0 \\ 0 & -1 & 1 & 0 & \cdots & 0 \\ \vdots & & & \ddots & & \vdots \\ 0 & \cdots & & 0 & -1 & 1 \end{bmatrix}.$$

The term $\boldsymbol{f}^{\mathrm{T}}H\boldsymbol{f}$ is proportional to $\int |f'(s)|^2\,ds$ and therefore gives preference to smoother solutions. The value of λ determines the trade-off between smoothness and agreement with Eq. (8.16). For $\lambda = 0$ (i.e., no regularization) the optimization problem (8.18) reduces to a least-squares formulation of Eqs. (8.16) and (8.17).

8.6 Examples

Numerical evidence suggests that each of the following delay differential equations possesses an invariant density.

Ershov's example: (with $\tau = 1/0.3$)

$$x(t)' = -x(t) + 1 - 1.9|x(t-\tau)| \tag{8.19}$$

Mackey–Glass equation:

$$x'(t) = -x(t) + \frac{2x(t-2)}{1 + \left(x(t-2)\right)^{10}} \tag{8.20}$$

Piecewise constant feedback:

$$x'(t) = -x(t) + \begin{cases} 4 & \text{if } 1 < x(t-6) < 2 \\ 0 & \text{otherwise.} \end{cases} \tag{8.21}$$

One can approximate these invariant densities empirically, by computing a histogram on an ensemble of numerical solutions of the differential delay equation. Figures 8.1, 8.2, and 8.3 show empirical invariant densities for each of the differential delay equations above, together with the empirical copula density $c(u, v)$ (cf. Eq. (8.13) that associates $f(x)$ with the invariant bivariate density $\rho(x, y)$).

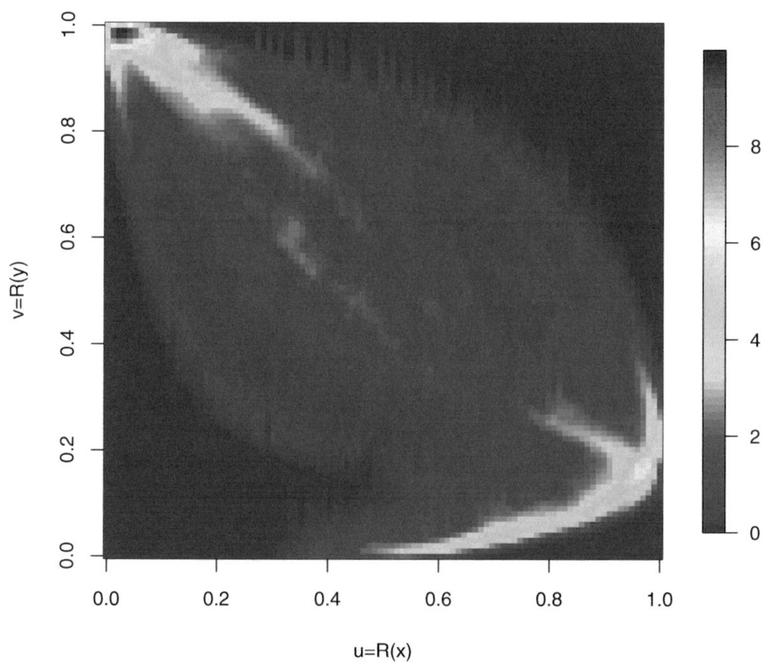

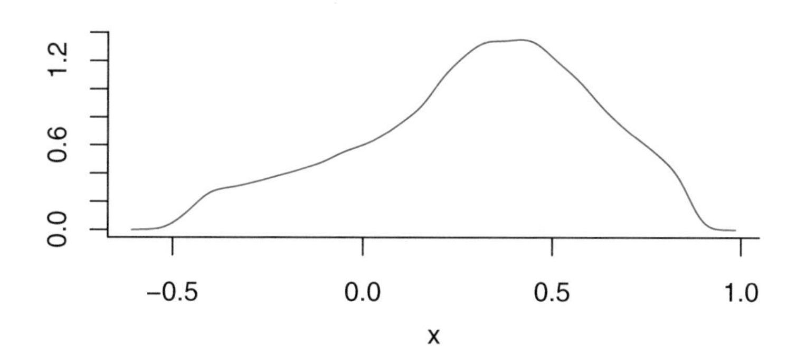

Fig. 8.1 Empirical invariant densities estimated by ensemble simulation, for Ershov's differential delay equation (8.19)

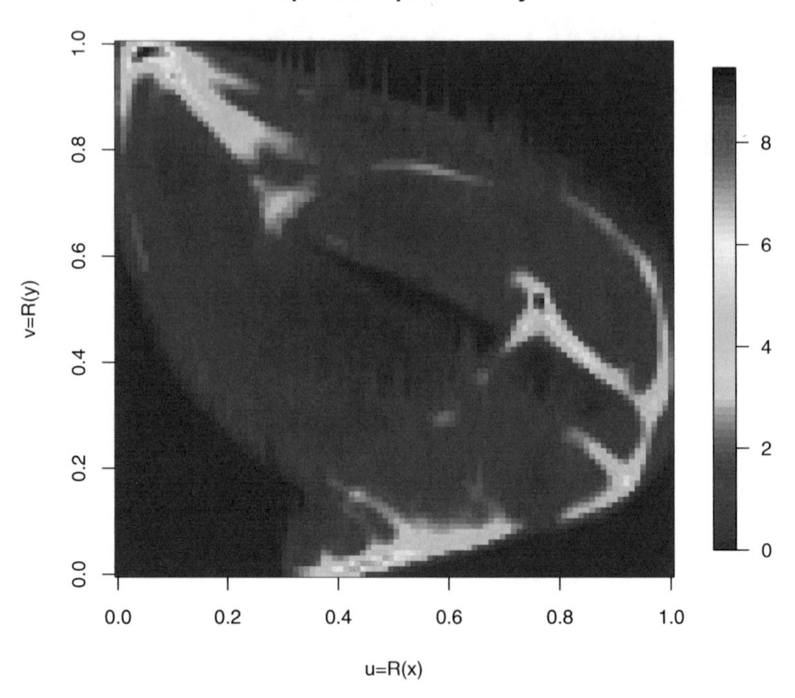

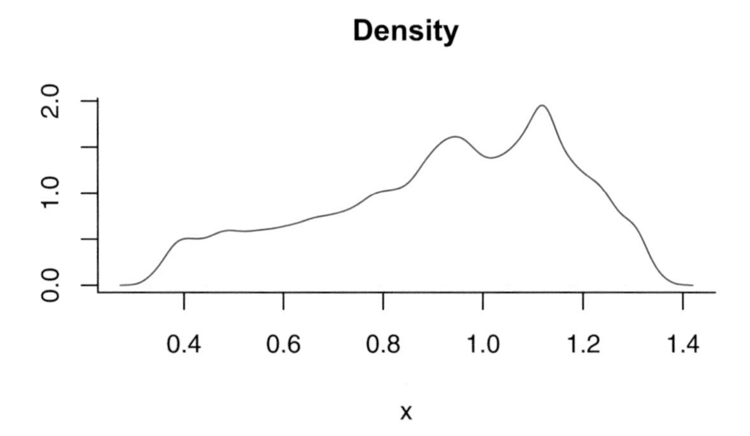

Fig. 8.2 Empirical invariant densities estimated by ensemble simulation, for the Mackey–Glass equation (8.20)

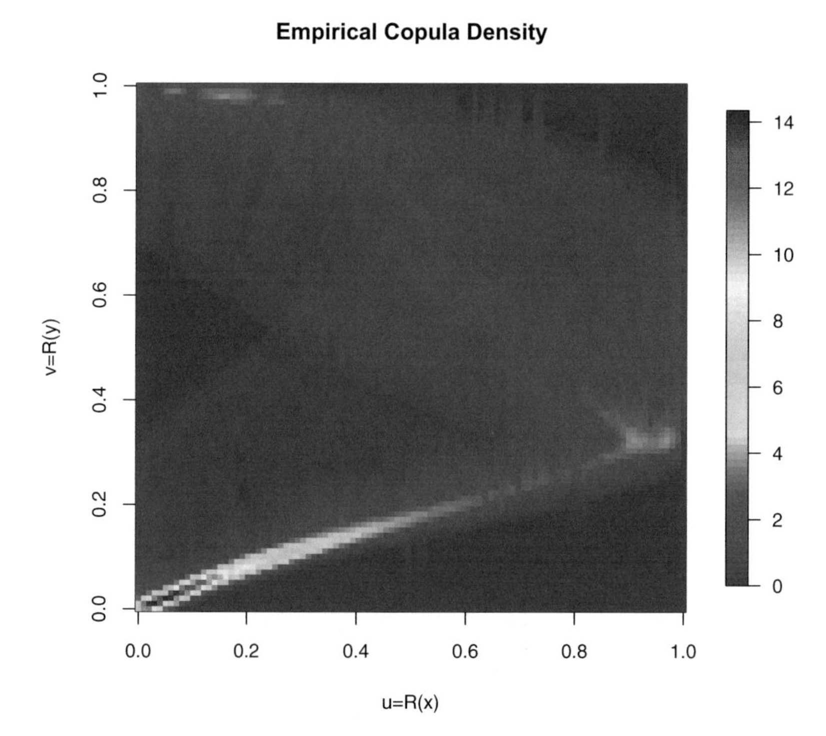

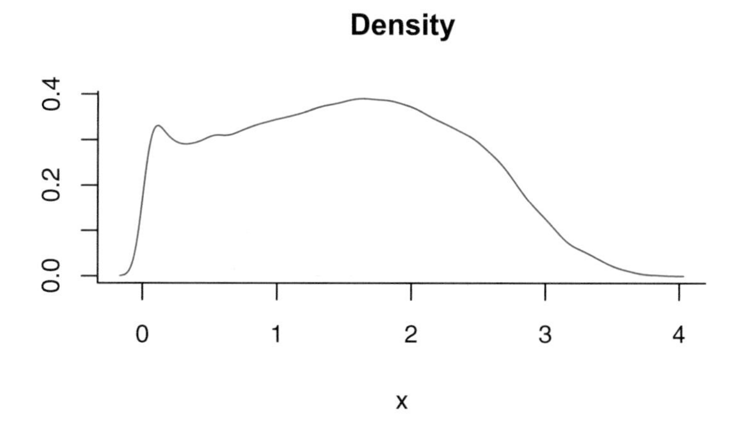

Fig. 8.3 Empirical invariant densities estimated by ensemble simulation, for the differential delay equation (8.21) with piecewise-constant feedback

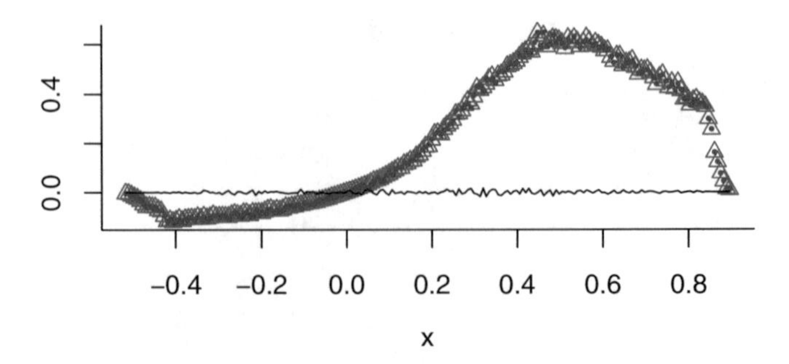

Fig. 8.4 Verification of Eq. (8.8) for the invariant density of Eq. (8.19) (Ershov's example). Open red triangles: left-hand side of Eq. (8.8). Solid blue dots: right-hand side of Eq. (8.8). Solid black line: residuals

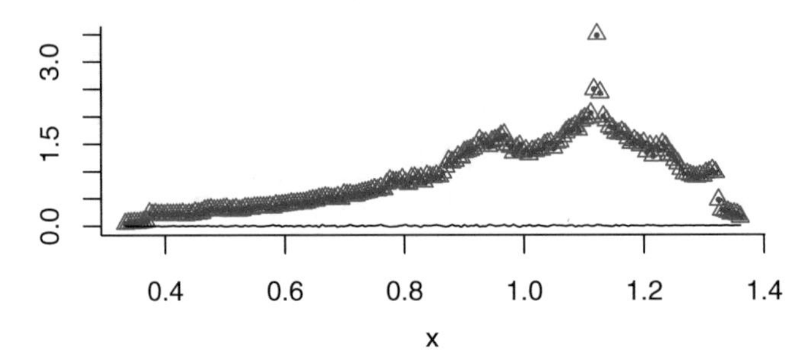

Fig. 8.5 Verification of Eq. (8.8) for the invariant density of Eq. (8.20) (Mackey–Glass equation). Open red triangles: left-hand side of Eq. (8.8). Solid blue dots: right-hand side of Eq. (8.8). Solid black line: residuals

Equation (8.8) is our main result characterizing invariant densities for delay differential equations. For each of the delay differential equations (8.19)–(8.21) we verify numerically that the empirical invariant density satisfies this relation. Figures 8.4, 8.5, and 8.6 show good agreement between the left- and right-hand sides of Eq. (8.8) for each of the delay differential equations considered. Any discrepancy can be accounted for by sampling error (due to finite sample size). Table 8.1 further demonstrates that the empirical densities agree (within sampling error) with Eq. (8.10).

The hope is that, for some copula density c, a solution of the optimization problem (8.18) will yield a good approximation of the invariant density for a given differential delay equation. However, without a priori knowledge of the true copula that gives the dependence structure of the invariant bivariate density, it is unclear how best to choose c from the infinite class of copulas available.

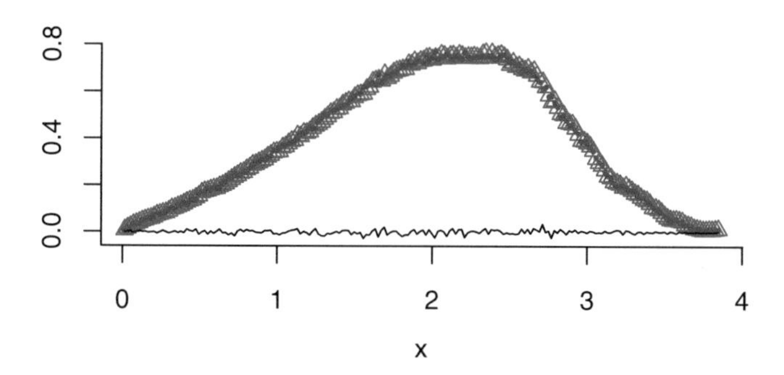

Fig. 8.6 Verification of Eq. (8.8) for the invariant density of Eq. (8.21) (differential delay equation with piecewise-constant feedback). Open red triangles: left-hand side of Eq. (8.8). Solid blue dots: right-hand side of Eq. (8.8). Solid black line: residuals

Table 8.1 Verification of relation (8.10) for the empirical invariant densities for the delay differential equations (8.19)–(8.21)

DDE	$\langle x \rangle$	$\langle S(x) \rangle$	Relative error
Ershov's example (8.19)	0.297895	0.297882	4×10^{-5}
Mackey–Glass (8.20)	0.91969	0.91972	3×10^{-5}
Piecewise constant (8.21)	1.4998	1.5030	2×10^{-3}

It is possible to recover the empirical invariant density by solving (8.18), provided the copula is known (that is, if we take c to be the empirical copula density). Here and in the following we have solved (8.18) with $n = 100$, using the optimization routine `spg` implemented in the `R` language. This routine requires an "initial guess" for the solution; we find that different initial guesses lead `spg` to converge to different local minima for (8.18). The empirical invariant density is recovered only with a very good initial guess. For Ershov's differential delay equation (8.19), Fig. 8.7 shows numerical solutions of (8.18) corresponding to different initial guesses, together with the empirical invariant density for comparison. In each case we have taken the regularization parameter to be $\lambda = 10^{-6}$.

If the correct copula is not known and we solve (8.18) using an arbitrary assumed copula density, varying results are obtained. The simplest choice is the uniform copula density $c(x, y) = 1$ (the "independence copula," which holds for any bivariate distribution in which x, y are independent). For each of the delay differential equations (8.19)–(8.21), Figs. 8.8, 8.9, and 8.10 show four numerical solutions of the optimization problem (8.18), corresponding to various combinations of assumed copula (independence and empirical) and initial guess (uniform and

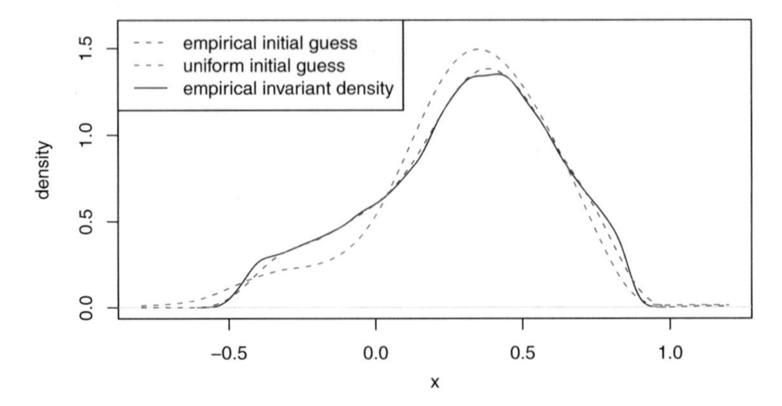

Fig. 8.7 Approximate invariant densities for Ershov's equation (8.19), obtained by numerical solution of the optimization problem (8.18) with different initial guesses as inputs to the optimization routine spg. The copula density c was taken to be the empirical copula as estimated by ensemble simulation. Also shown for comparison is the invariant density as estimated by ensemble simulation

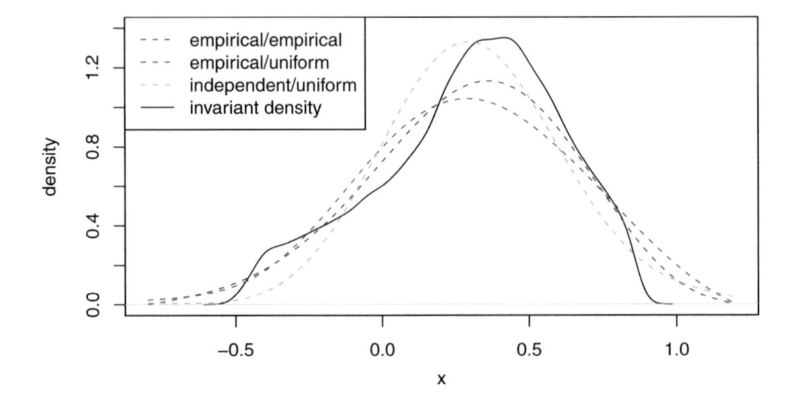

Fig. 8.8 Approximate invariant densities for Ershov's equation (8.19), obtained by numerical solution of the optimization problem (8.18) using a variety of combinations of assumed copula and initial guess as inputs to the optimization routine spg. Legend text indicates [form of assumed copula density]/[form of initial guess for invariant density]. Also shown for comparison is the invariant density as estimated by ensemble simulation

empirical). In each case we have taken $\lambda = 10^{-5}$. As Figs. 8.8, 8.9, and 8.10 show, in most cases the numerical solution of the optimization problem gives a fairly poor estimate of the invariant density, unless the correct copula and a good initial guess for the invariant density are used. However, even for poor choices of copula and initial guess, the numerical solutions in almost all cases give a remarkably good estimate for the *support* of the invariant density.

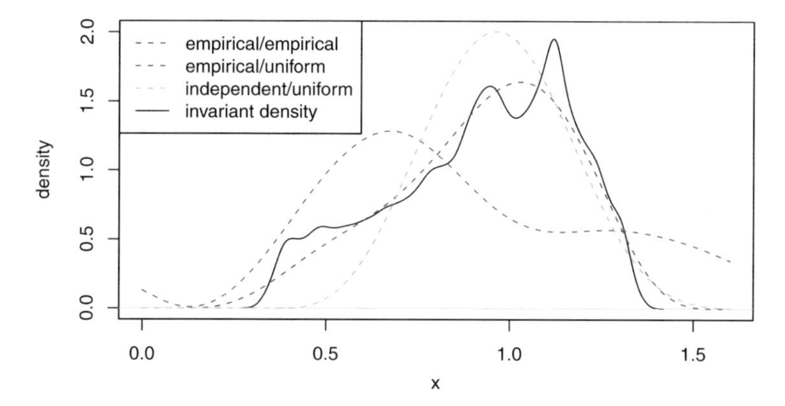

Fig. 8.9 Approximate invariant densities for the Mackey–Glass equation (8.20), obtained by numerical solution of the optimization problem (8.18) using a variety of combinations of assumed copula and initial guess as inputs to the optimization routine `spg`. Legend text indicates [form of assumed copula density]/[form of initial guess for invariant density]. Also shown for comparison is the invariant density as estimated by ensemble simulation

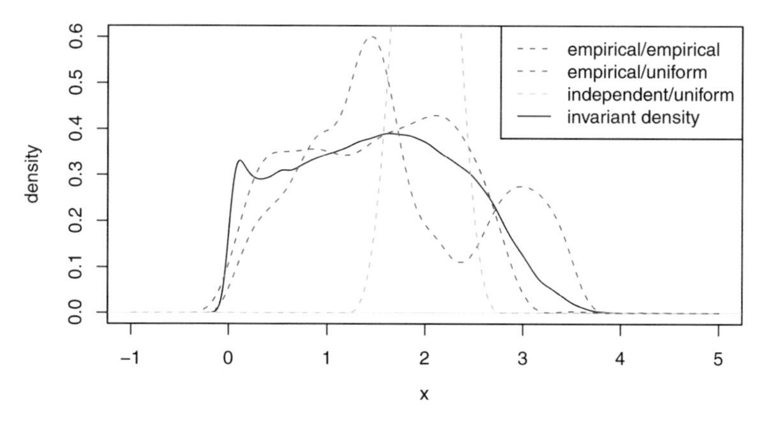

Fig. 8.10 Approximate invariant densities for the differential delay equation (8.21) with piecewise-constant feedback, obtained by numerical solution of the optimization problem (8.18) using a variety of combinations of assumed copula and initial guess as inputs to the optimization routine `spg`. Legend text indicates [form of assumed copula density]/[form of initial guess for invariant density]. Also shown for comparison is the invariant density as estimated by ensemble simulation

The regularization parameter λ has a significant influence on the agreement between the empirical invariant density and the numerical solution of (8.18), especially when the correct copula is unknown. For Ershov's equation (8.19), Fig. 8.11 shows four numerical solutions of (8.18) corresponding to different values of λ. Here we have assumed the independence copula and taken a uniform density as the initial guess, as inputs to the optimization routine `spg`.

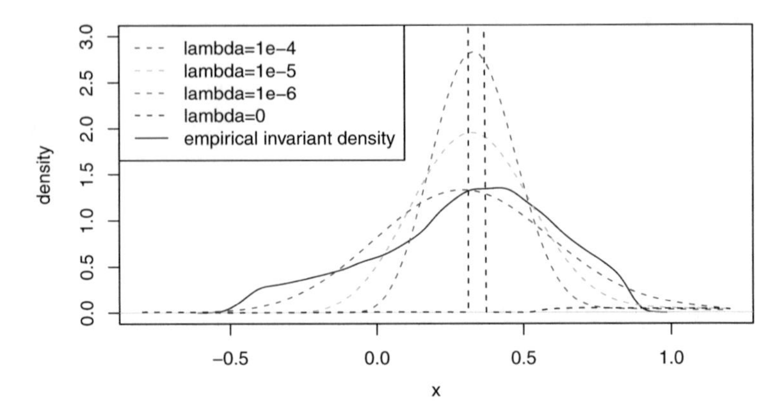

Fig. 8.11 Approximate invariant densities for Ershov's equation (8.19), obtained by numerical solution of the optimization problem (8.18) for different values of the regularization parameter λ. The copula density c was taken to be the empirical copula as estimated by ensemble simulation, and a uniform density was taken as the initial guess in the optimization routine `spg`. Also shown for comparison is the invariant density as estimated by ensemble simulation

Chapter 9
Summary and Conclusions

This is a weird monograph to say the least. Rather than presenting a body of finished work with theorems and proofs and examples and applications, we have presented the problem of how to treat the evolution of densities under that action of delayed dynamics and given no solutions! We have simply illustrated the problem in Chaps. 1–3 and then given all of the reasons why it is difficult mathematically in Chap. 4. Following this we have presented a series of chapters in which we detail various attempts that have been made to solve the problem, all of which have led to naught. This is definitely NOT your standard mathematical monograph! So why have we done this? Simply to lay out a map of what we think are blind alleys for the unwary neophyte starting out in search of a solution.

In our collective opinion, based on years of fruitless attempts, the most promising avenue seems to be that of the Hopf functionals in Chap. 5, based on the work of Capiński (1991), but we have not taken it further than what is detailed here. The second most promising may be that of Sect. 7.2 in which we present analytic results based on the work of Tulcea and Marinescu (1950), but it is unclear how to extend these results. If our suspicions about either of these two approaches turn out to bear fruit then we will be gratified and applaud the appearance of concrete progress.

One aspect that we have *not* written about is that of stochastic differential delay equations. There is an extensive literature dealing with this, and in our opinion none of it deals effectively with the substantiative mathematical issues raised in Chap. 4. The main paper of interest in this literature is that of Guillouzic et al. (1999) and the review Longtin (2009).

J. Losson et al., *Density Evolution Under Delayed Dynamics*, Fields Institute Monographs 38, https://doi.org/10.1007/978-1-0716-1072-5_9

References

Blythe S, Nisbet R, Gurney W (1984) The dynamics of population models with distributed maturation periods. Theoret Popul Biol 25(3):289–311

Brunovsky P, Komornik J (1984) Ergodicity and exactness of the shift on $C[0, \infty)$ and the semiflow of a first-order partial differential equation. J Math Anal Appl 104:235–245

Capiński M (1991) Hopf equation for some nonlinear differential delay equation and invariant measures for corresponding dynamical system. Univ Iagel Acta Math (28):171–175

Carrier GF, Pearson CE (1988) Partial differential equations, 2nd edn. Academic Press, San Diego

Collet P, Eckmann JP (1980) Iterated Maps of the Interval as Dynamical Systems. Birkhäuser, Cambridge

Cooke KL, Krumme DW (1968) Differential-difference equations and nonlinear initial-boundary value problems for linear hyperbolic partial differential equations. J Math Anal Appl 24(2):372–387

Davis MHA (1984) Piecewise-deterministic Markov processes: a general class of nondiffusion stochastic models. J Roy Statist Soc Ser B 46(3):353–388, with discussion

Diekmann O, van Gils SA, Lunel SMV, Walther HO (1995) Delay Equations: Functional-, Complex-, and Nonlinear Analysis, Applied Mathematical Sciences, vol 110. Springer-Verlag, New York

Dorrizzi B, Grammaticos B, Berre ML, Pomeau Y, Ressayre E, Tallet A (1987) Statistics and dimension of chaos in differential delay systems. Phys Rev A 35(1):328–339

Driver RD (1977) Ordinary and Delay Differential Equations, vol 20. Springer Science & Business Media

Ershov SV (1991) Asymptotic theory of multidimensional chaos. J Stat Phys 69(3/4):781–812

Fargue D (1973) Réducibilité des systèmes héréditaires à des systèmes dynamiques. C R Acad Sci Paris 277B:471, (French)

Fargue D (1974) Réducibilité des systèmes héréditaires. Int J Nonlin Mech 9:331, (French)

Foias C, Prodi G (1976) Sur les solutions statistiques des équations de Navier-Stokes. Ann Mat Pura Appl 111(1):307–330

Gardiner C (1983) Handbook of Stochastic Methods. Springer Verlag, Berlin, Heidelberg

Gardiner C (1991) Quantum Noise. Springer Verlag, Berlin, Heidelberg

Genest C, Mackay J (1986) The joy of copulas: Bivariate distributions with uniform marginals. Amer Statist 40(4):280–283

Giusti E, Williams GH (1984) Minimal Surfaces and Functions of Bounded Variation, vol 2. Springer

Gora P, Boyarsky A (1989) Absolutely continuous invariant measures for piecewise expanding C^2 transformations in R^n. Israel J Math 67(3):272–286

J. Losson et al., *Density Evolution Under Delayed Dynamics*, Fields Institute Monographs 38, https://doi.org/10.1007/978-1-0716-1072-5

Guillouzic S, L'Heureux I, Longtin A (1999) Small delay approximation of stochastic delay differential equations. Phys Rev E 61:3970–3982

Gwizdz P, Tyran-Kamińska M (2019) Densities for piecewise deterministic Markov processes with boundary. J Math Anal Appl 479(1):384–425

Hale JK, Lunel SMV (2013) Introduction to Functional Differential Equations, vol 99. Springer Science & Business Media

an der Heiden U (1985) Stochastic properties of simple differential-delay equations. In: Meinardus G, Nurnberger G (eds) Delay Equations, Approximation and Application, International Series of Numerical Mathematics, vol 74, Birkhäuser, pp 147–164

an der Heiden U, Mackey MC (1982) The dynamics of production and destruction: Analytic insight into complex behavior. J Math Biol 16:75–101

Hopf E (1952) Statistical hydromechanics and functional calculus. J Rat Mech Anal 1:87–123

Hunt BR, Kennedy JA, Li TY, Nusse HE (2002) SLYRB measures: natural invariant measures for chaotic systems. Phys D 170(1):50–71

Ito S, Tanaka S, Nakada H (1979a) On unimodal linear transformations and chaos. I. Tokyo J Math 2(2):221–239

Ito S, Tanaka S, Nakada H (1979b) On unimodal linear transformations and chaos. II. Tokyo J Math 2(2):241–259

Ivanov AF, Šarkovskiĭ AN (1991) Oscillations in singularly perturbed delay equations. Dynamics Reported 3:165, Springer Verlag (edited by H. O. Walter and U. Kirchgraber)

Kac M (1980) Integration in Function Spaces and some of its Applications. Accademia Nazionale dei Lincei, Scuola Normale Superiore, Pisa

Keller G, Künzle M (1992) Transfer operators for coupled map lattices. Ergodic Theory Dynam Syst 12(2):297–318

Komorník J (1993) Asymptotic periodicity of Markov and related operators. In: Dynamics reported, Dynam. Report. Expositions Dynam. Systems (N.S.), vol 2, Springer, Berlin, pp 31–68

Komorník J, Lasota A (1987) Asymptotic decomposition of Markov operators. Bull Polish Acad Sci Math 35(5-6):321–327

Kuznetsov YA (1995) Elements of Applied Bifurcation Theory, Applied Mathematical Sciences, vol 112. Springer-Verlag, New York

Lang S (1993) Real and Functional Analysis, Graduate Texts in Mathematics, vol 142, 3rd edn. Springer-Verlag, New York—Berlin—Heidelberg

Lasota A, Mackey MC (1987) Noise and statistical periodicity. Phys D 28(1-2):143–154

Lasota A, Mackey MC (1994) Chaos, Fractals, and Noise: Stochastic Aspects of Dynamics, Applied Mathematical Sciences, vol 97. Springer-Verlag, New York

Lepri S, Giacomelli G, Politi A, Arecchi FT (1993) High-dimensional chaos in delayed dynamical systems. Phys D 70:235–249

Lewis RM, Kraichnan RH (1962) A space-time functional formalism for turbulence. Commun Pure Appl Math 15(4):397–411

Longtin A (2009) Stochastic delay-differential equations. In: Complex time-delay systems, Springer, pp 177–195

Lorenz EN (1963) Deterministic non-periodic flow. J Atmos Sci 20:130–141

Losson J (1991) Multistability and probabilistic properties of differential delay equations. Montréal, Quebec, CANADA

Losson J, Mackey MC (1992) A Hopf-like equation and perturbation theory for differential delay equations. J Stat Phys 69(5/6):1025–1046

Losson J, Mackey MC (1995) Coupled map lattices as models of deterministic and stochastic differential delay equations. Phys Rev E 52(1):115–128

Losson J, Milton J, Mackey MC (1995) Phase transitions in networks of chaotic elements with short and long range interactions. Phys D 81(1-2):177–203

Mackey MC, Glass L (1977) Oscillation and chaos in physiological control systems. Science 197:287–289

Milton JG, Mackey MC (2000) Neural ensemble coding and statistical periodicity: Speculations on the operation of the mind's eye. J Physiol Paris 94(5–6):489–503

Mitkowski P (2021) Mathematical Structures of Ergodicity and Chaos in Population Dynamics. Studies in Systems, Decision and Control, vol 312. Springer, Cham

Mitkowski P, Mitkowski W (2012) Ergodic theory approach to chaos: Remarks and computational aspects. Int J Appl Math Comput Sci 22(2):259–267

Muir T, Metzler WH (2003) A Treatise on the Theory of Determinants. Courier Corporation

Nelsen RB (2006) An Introduction to Copulas, 2nd edn. Springer

Press WH, Teukolsky SA, Vetterling WT, Flannery BP (1992) Numerical Recipes in C, 2nd edn. Cambridge University Press

Provatas N, Mackey MC (1991a) Asymptotic periodicity and banded chaos. Phys D 53(2-4):295–318

Provatas N, Mackey MC (1991b) Noise-induced asymptotic periodicity in a piecewise linear map. J Statist Phys 63(3-4):585–612

Risken H (1984) The Fokker-Planck Equation. Springer-Verlag, Berlin, New York, Heidelberg

Rudnicki R (1985) Invariant measures for the flow of a first order partial differential equation. Ergodic Theory Dynam Syst 5:437–443

Rudnicki R (1987) An abstract Wiener measure invariant under a partial differential equation. Bull Polish Acad Sci Math 35(5–6):289–295

Rudnicki R (1988) Strong ergodic properties of a first-order partial differential equation. J Math Anal Appl 133:14–26

Rudnicki R (1993) Gaussian measure-preserving linear transformations. Univ Iagel Acta Math (30):105–112

Rudnicki R (2015) An ergodic theory approach to chaos. Discrete Contin Dynam Syst Ser A 35(2):757

Rudnicki R, Tyran-Kamińska M (2017) Piecewise Deterministic Processes in Biological Models. Springer Briefs in Applied Sciences and Technology, Springer, Cham

Ryder LH (1985) Quantum Field Theory. Cambridge University Press, Cambridge

Sander E, Barreto E, Schiff S, So P (2005) Dynamics of noninvertibility in delay equations. Discrete Contin Dynam Syst 2005:768–777

Shampine LF, Thompson S (2001) Solving DDEs in MATLAB. Appl Numer Math 37(4):441–458, URL www.runet.edu/~thompson/webdes/

Sharkovskiĭ AN, Maĭstrenko YL, Romanenko EY (1986) Raznostnye uravneniya i ikh prilozheniya. "Naukova Dumka", Kiev

Sklar A (1959) Fonctions de répartition à n dimensions et leurs marges. Publ Inst Statist Univ Paris 8:229–231

Sobczyk K (1984) Stochastic Wave Propagation. PWN—Polish Scientific Publishers, Warsaw, translated from the Polish by the author, I. Bychowska and Z. Adamowicz

Taylor SR (2004) Probabilistic properties of delay differential equations. Waterloo, Ontario, Canada, arXiv:190902544

Taylor SR (2011) Liouville-like equations and invariant densities for delay differential equations

Tulcea CI, Marinescu G (1950) Théorie ergodique pour des classes d'opérations non complètement continues. Ann Math 52(1):140–147

Vogel T (1965) Théorie des systèmes évolutifs, vol 22. Gauthier-Villars

Yamasaki Y (1985) Measures on Infinite Dimensional Spaces, Series in Pure Mathematics, vol 5. World Scientific, Singapore—Philadelphia

Yoshida T, Mori H, Shigematsu H (1983) Analytic study of chaos of the tent map: band structures, power spectra, and critical behaviors. J Statist Phys 31(2):279–308

Young LS (2002) What are SRB measures, and which dynamical systems have them? J Stat Phys 108:733–754

Zauderer E (1983) Partial Differential Equations of Applied Mathematics. Wiley, New York

Ziemer WP (2012) Weakly Differentiable Functions: Sobolev Spaces and Functions of Bounded Variation, vol 120. Springer Science & Business Media

Index

J. Losson et al., *Density Evolution Under Delayed Dynamics*, Fields Institute Monographs 38, https://doi.org/10.1007/978-1-0716-1072-5

Printed in the United States
by Baker & Taylor Publisher Services